IMMUNOLOGY AND IMMUNE SYSTEM DISORDERS

# OLD AND NOVEL HUMORAL BIOMARKERS OF AUTOIMMUNE MYASTHENIA GRAVIS

# IMMUNOLOGY AND IMMUNE SYSTEM DISORDERS

Additional books in this series can be found on Nova's website under the Series tab.

Additional e-books in this series can be found on Nova's website under e-books tab.

IMMUNOLOGY AND IMMUNE SYSTEM DISORDERS

# OLD AND NOVEL HUMORAL BIOMARKERS OF AUTOIMMUNE MYASTHENIA GRAVIS

DAVIDE GIACOMO CORDA

GIOVANNI ANDREA DEIANA

GIANNINA ARRU

GIOVANNI MASALA

AND

GIANPIETRO SECHI

nova

Medicine & Health

New York

We have partnered with Copyright Clearance Center to make it easy for you to obtain permissions to reuse content from this publication. Simply navigate to this publication's page on Nova's website and locate the "Get Permission" button below the title description. This button is linked directly to the title's permission page on copyright.com. Alternatively, you can visit copyright.com and search by title, ISBN, or ISSN.

For further questions about using the service on copyright.com, please contact:
Copyright Clearance Center
Phone: +1-(978) 750-8400          Fax: +1-(978) 750-4470          E-mail: info@copyright.com.

## NOTICE TO THE READER

The Publisher has taken reasonable care in the preparation of this book, but makes no expressed or implied warranty of any kind and assumes no responsibility for any errors or omissions. No liability is assumed for incidental or consequential damages in connection with or arising out of information contained in this book. The Publisher shall not be liable for any special, consequential, or exemplary damages resulting, in whole or in part, from the readers' use of, or reliance upon, this material. Any parts of this book based on government reports are so indicated and copyright is claimed for those parts to the extent applicable to compilations of such works.

Independent verification should be sought for any data, advice or recommendations contained in this book. In addition, no responsibility is assumed by the publisher for any injury and/or damage to persons or property arising from any methods, products, instructions, ideas or otherwise contained in this publication.

This publication is designed to provide accurate and authoritative information with regard to the subject matter covered herein. It is sold with the clear understanding that the Publisher is not engaged in rendering legal or any other professional services. If legal or any other expert assistance is required, the services of a competent person should be sought. FROM A DECLARATION OF PARTICIPANTS JOINTLY ADOPTED BY A COMMITTEE OF THE AMERICAN BAR ASSOCIATION AND A COMMITTEE OF PUBLISHERS.

Additional color graphics may be available in the e-book version of this book.

## Library of Congress Cataloging-in-Publication Data

ISBN: 978-1-53613-836-8

*Published by Nova Science Publishers, Inc. † New York*

# CONTENTS

# *Contents*

# PREFACE

Autoimmune myasthenia gravis (MG) is mediated by autoantibodies to components of the postsynaptic muscle endplate at the neuromuscular junction. Due to the clinical heterogeneity of the disease, there is a great need for reliable biomarkers useful for diagnostic as well as therapeutic purposes. Humoral biomarkers of MG can be divided into two categories: 1) autoantibodies; 2) other immune-related molecules such as inflammatory proteins, microRNA, HLA genes.

Regarding autoantibodies, IgG1 and IgG3 antibodies as has been revealed by radio immune assay (RIA), are directed against the nicotinic acetylcholine receptors (AChRAb) in 85% of the patients with generalized MG but only in 50% of those with the ocular form. In these forms, thymoma or thymic hyperplasia can occur. In AChRAb negative generalized MG patients, a variable percentage of patients had IgG4 antibodies towards muscle specific tyrosine kinase (MuSKAb). More rarely, IgG towards low density lipoprotein receptor-related protein 4 (Lrp4Ab), agrin or cortactin may be detected. In many RIA seronegative patients, the more sensitive cell based assay method can detect the above-mentioned antibodies. Finally, titin and ryanodine receptor antibodies, named striational antibodies, can occur in association in patients with AChRAb positive MG indicating either the possibility of thymoma or, in the context of late-onset myasthenia gravis, the occurrence of a severe form of the disease which shows "no response" to thymectomy and needs long-term immunosuppression.

As regards immune-related molecules, recent studies have reported increased serum levels of a proliferation-inducing ligand (APRIL), some cytokines, matrix metalloproteinase 10 (MMP-10), transforming growth factor alpha and the extracellular newly identified receptor for advanced glycation end-products binding protein. These last three proteins are involved in the cell cycle progression and differentiation and play multiple roles in the immune response. Moreover, circulating microRNAs (miRNA) have been reported to be potential biomarkers in AChRAb positive MG and in MuSKAb positive MG. The miRNAs are involved in the development of T- and B-cells autoimmune response. Some HLA associations have been reported, such as DR3-B8-A1 in early onset AChRAb MG and DR14, DR16 and DQ5 in MuSKAb MG.

Autoantibodies against AChR, MuSK or LRP4 protein and striational antibodies are MG biomarkers currently used in clinical practice. Other immune-related molecules have not yet a clinical use but may give insights into the pathogenesis of MG and possibly will have some clinical utility in the future.

*Chapter 1*

# INTRODUCTION

## 1.1. AUTOIMMUNE MYASTHENIA GRAVIS

Autoimmune myasthenia gravis (MG) is an acquired disorder of neuromuscular transmission mediated by antibodies against nicotinic acetylcholine receptors (AChRAb) or against other proteins of the neuromuscular junction (NMJ) (Figure 1; Figure 2; Figure 3). Based on 55 studies spanning 1950–2007, the incidence is 5.3 per million person-years and the prevalence is 77.7 cases per million [1, 2] but rates are increasing particularly in older individuals. There is a bimodal peak of incidence with a prevalence of female patients under the age of 40 and of males over the age of 50 [3].

The hallmark of MG is the impairment of neuromuscular transmission. Patients experience weakness with fatigability on skeletal muscles, predominantly facial, bulbar and proximal muscles of limbs. A simple clinical classification distinguishes pure ocular MG from the generalized form with mild, moderate or severe manifestation. In ocular myasthenia only the outer ocular muscles including the levator palpebrae are affected, nearly always asymmetrically, with ptosis and double vision. These symptoms may be transient, fluctuating or progressive during the day. About 10–20% of patients continue to have a pure ocular MG. The majority of patients proceed to generalized and symmetrical muscle fatigability and weakness within 24 months after disease onset [4]. Generalized MG consists of any clinical impairment of muscle groups other than outer ocular muscles independent of its severity. This generic description corresponds mainly to the common AChRAb seropositive form of the disease.

Diagnosis of MG is based on evidenced weakness with fatigability of skeletal muscles. If MG is clinically suspected, confirmation by autoantibody testing for AChRAb is appropriate. If these antibodies are negative, other autoantibodies such as MuSKAb can be checked. Electro-diagnostic testing with repetitive nerve stimulation or, if negative, single-fiber electromyography, is helpful in seronegative cases or in urgent and emergency care [3]. In doubtful cases, an unequivocal clinical response to acetylcholinesterase inhibitors may be of some benefit. When MG is diagnosed, chest imaging should be performed to evidence thymoma or thymic hyperplasia. Therapeutic options include acetylcholinesterase inhibitors, immuno-suppressive treatment, plasma exchange and intravenous immunoglobulins (IVIG), thymectomy and supportive treatment depending on the peculiar features of the single patient [5].

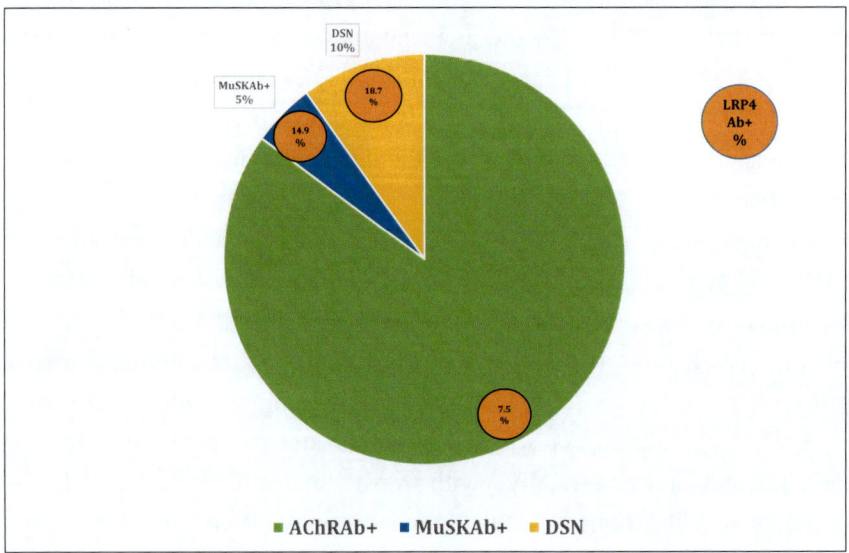

Figure 1. Main autoimmune subgroups in Myasthenia Gravis patients from the literature and percentage of LRP4Ab positivity in these subgroups according to ref. 197.

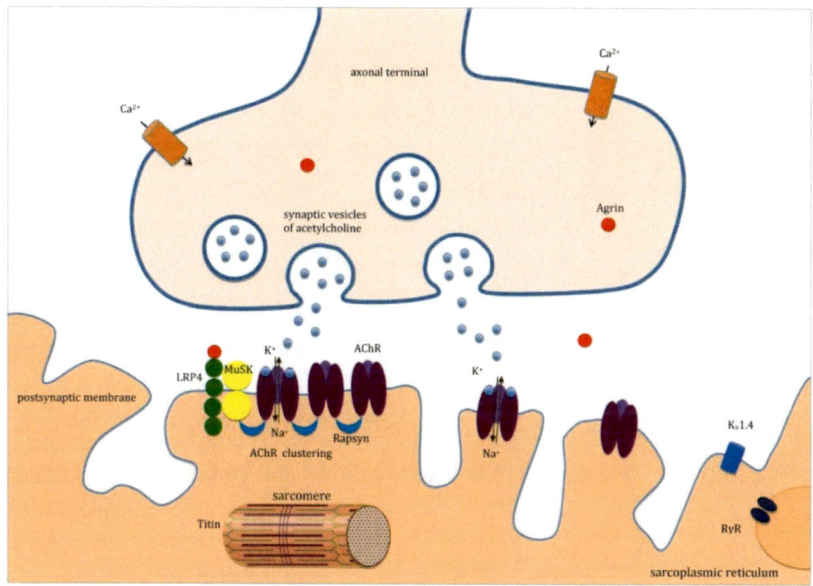

Figure 2. Normal neuromuscular junction with its main antigenic targets.

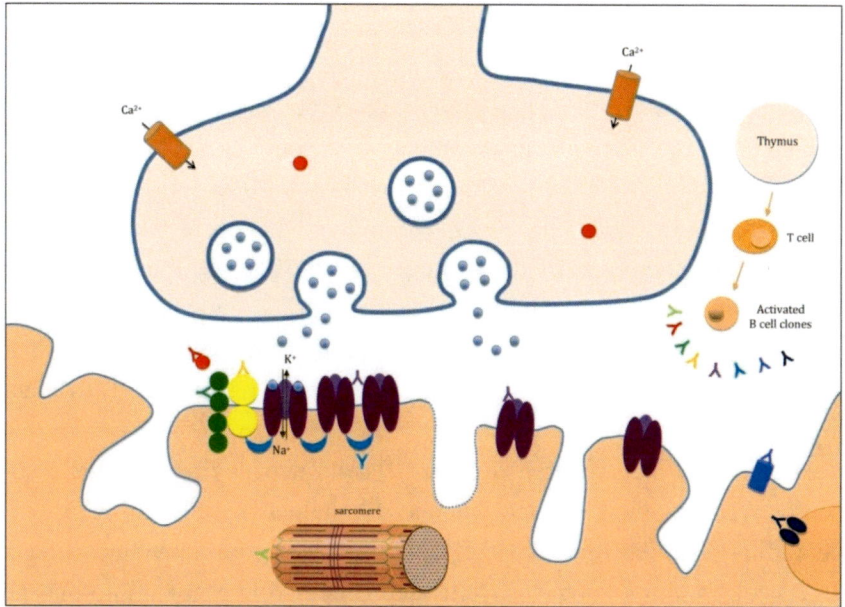

Figure 3. Antibody attack on neuromuscular junction.

## 1.2. PATHOGENESIS OF MG: IMMUNE SYSTEM

Pathogenesis of autoimmune MG is only partially understood because of the imperfect knowledge of the immune system.

The immune system is a host defense system that can distinguish pathogens' agents from self-tissues. In humans (and vertebrates) the immune system can be divided into an innate and adaptive immune system or in humoral and cell-mediated immunity [6].

The innate immune system provides an immediate and non-specific response to pathogens. If this response is unsatisfactory, an adaptive immune system is activated that can adapt and enhance its response also retaining it as immune memory (long lasting immunity). The cells of the adaptive immune system are lymphocytes classified into B cells, engaged in humoral immunity, and T cells, involved in cell-mediated immunity.

Both T-cell and B-cells originate from stem cells in the bone marrow. Whereas B cells also undergo their maturation process in the bone marrow, T-cells need to migrate to the thymus where they mature. Bone marrow and thymus are primary lymphoid tissues.

Both B cells and T cells have receptors that recognize a "non-self" antigen. T cells are divided into killer T, helper T and regulatory T cells all needing the antigen processing, namely the presentation of an antigen in combination with the major histocompatibility complex (MHC) molecule. Killer T cells identify antigens bound to Class I MHC molecules, while helper and regulatory T cells recognize antigens bound to Class II MHC molecules. Another type of recognition cells are the $\gamma\delta$ T cells that can identify antigens not bound to MHC receptors [7]. B cells have antibodies on their surface directly permitting the recognition of antigens without antigen processing. Each lineage of B and T cells recognizes a different antigen.

The process of activation of killer T cells depends partly on the co-receptor CD8 on their surface whereas for helper T cells, the CD4 co-receptor has a similar role. Helper T cells collaborate for activating killer T cells, B cells and macrophages. Activated B cells differentiate into plasma cells that secrete large volumes of antibodies. T and B cells also become memory cells.

Autoimmunity is a disorder of the immune system that does not distinguish between self and non-self or foreign antigens and consequently attacks its own tissues. Normally, many T cells and antibodies match self-antigens but are eliminated by specialized organs such as the thymus and bone marrow [8, 9].

## 1.3. PATHOGENESIS OF MG: THYMUS

In the pathogenesis of MG, at least in the AChRAb positive form, thymus seems to have an important role.

The thymus is a primary lymphoid organ of the immune system with endocrine functions, composed of two lobes and located in the anterior superior mediastinum. The thymus grows from birth to puberty and then begins to shrink with a replacement of the organ with fat, in connection with the effects of sex hormones; this thymic involution appears to be directed by the high levels of circulating hormones [10]. Simultaneously, thymic size and its activity is maximal before puberty, then reduces in adult and old life.

At the anatomo-histologic level, the thymus can be divided into central medulla and peripheral cortex enclosed in a capsule. Thymic cells are stromal cells and cells derived from bone marrow hemato-poietic stem cells. Thymic stromal cells include epithelial and dendritic cells [11], mesenchymal cells [12], and myoid cells [13].

Thymic myoid cells express functional AChRs (both adult and fetal forms) consisting of folded subunits, epithelial cells express unfolded AChR subunits (but not whole functional channels) [4]. Both thymic epithelial cells and myoid cells have been hypothesized to generate autoimmunity in MG [14].

The so-called thymocytes are T cells in maturation, derived from bone marrow hematopoietic stem cells. In physiologic conditions B cells are very few [15].

The thymus performs the development of mature T cells from hematopoietic progenitor cells. Actually, in the cortex T cell receptor (TCR), gene rearrangement and positive selection are achieved. The process of rearrangement of TCR consists of the so-called V(D)J (variable, diversity, joining gene segments) recombination, a nearly random mixing of gene segments, obtaining a large number of possible combinations for the possible antigens: at maturation, each lymphocyte will have a specific TCR. Positive selection is intended for eliminating, by apoptosis, limphocytes provided with dysfunctional TCR without interaction (or with a weak interaction) with self MHC molecules of thymic epithelial cells. Negative selection is realized in the interaction of limphocytes with thymic epithelial cells and dendritic cells that express self antigens: in case of high affinity interaction, autoreactive T cells are destroyed through apoptosis: this process begins in the cortex and is completed in the medulla [16]. Negative selection is based on the correct functioning of the Autoimmune Regulator (AIRE) gene [17]. AIRE permits the expression of antigens of other tissues in the medulla allowing for the recognition (and consequently deletion) of autoreactive thymocytes realizing the so-called central tolerance. The possibility exists that some T lymphocytes escape from the negative selection of the thymus and go into the bloodstream: in such a case, peripheral mechanisms of silencing can operate. These mechanisms include clonal anergy, clonal deletion and immunoregulation by regulatory T cells [18].

Clonal anergy is consequent to the activation of TCR by an antigen without activation of the costimulatory receptor CD28. This happens in some cases e. g. when T cells interact with an antigen not presented by antigen presenting cells or if costimulatory molecules are downregulated on T cell. In these cases, in T cells there is the activation only of one transcriptional factor, NFAT, which translocates to the lymphocyte nucleus. Without the co-stimulation of CD28, there is no activation of other transcription factors necessary, with NFAT, for the induction of genes for the productive response of T cells [19, 20]. Consequently, NFAT homodimerizes and induces genes for anergy [21]. The T cell is activated to a productive response when the antigen is properly displayed on MHC II complex and the activation of co-receptors is performed. Peripheral tolerance is divided into recessive (as described above) and dominant tolerances. In this latter case, T regulatory cells (T-reg) actively inhibit the immune response; also in this case, NFAT signaling is involved. Tregs express the biomarkers CD4, CD25 and FOXP3 and are developed either in the thymus during the negative selection or in the immune periphery by conversion of autoreactive T cells (see also paragraph about T cell biomarkers) [22]. In particular, B and T cells are controlled in the periphery for their reactivity to self antigens by dendritic cells and lymph node stromal cells; if autoreactive, lymphocytes receive an apoptotic signal (clonal deletion) or can also be converted to Treg.

In addition, lymphocytes receive signals from female sex hormones (immunostimulation that explains the high prevalence of autoimmune diseases in women) [23] and male sex hormones (they seem to be immunosuppressive) [24], vitamin D. T cell necessitates vitamin D in its active hormone form, calcitriol, to perform its function; actually, T cells have a receptor for calcitriol, and a gene (CYP27B1) to convert calcidiol into calcitriol [25, 26]. Moreover, they show several other receptors, including those for prolactin and growth hormone [27, 28].

When mechanisms of tolerance fail, MG can develop.

In the AChRAb MG patients, the thymus displays structural and functional changes in at least 75% of cases [29]. Developing B cells infiltrate into germinal centers (although the thymus may not be grossly enlarged) forming follicles (follicular hyperplasia) or distributed throughout the thymic medulla (diffuse hyperplasia or thymitis) [30]; thymoma also can develop [31, 32]. Lymphofollicular hyperplasia is shown in 70% of patients affected by early onset MG. Unknown environmental etiologic factor(s) work in the context of genetic predisposition to determine lymphofollicular hyperplasia: among others, Epstein Barr virus [30] and human polyomavirus 7 [33] have been considered in AChRAb MG. The initial etiologic factor stimulates thymic epithelial cells (TECs) that become hyperplastic and present unfolded AChR subunits with MHC II to CD4+ T cells. These latter cells cause an initial antibody production against nearby myoid cells that express folded

AChRs. Thus, the complement is activated and AChR/immune complexes are released. These immune molecules stimulate professional antigen presenting cells that activating T and consequently B autoreactive cells can diversify antibody production against the native AChR [34]. Dysfunctional regulatory T cells would intervene to perpetuate the autoimmune process in the thymus. Successively, the autoimmune process spreads to the peripheral lymphatic tissue where it can continue also with the contribution of skeletal muscle derived AChR/immune complexes and dysfunctional regulatory T cells [4, 35].

Thymomas are present in 10-15% of patients with myasthenia gravis among Caucasians and up to 30% in individuals of Asian origin [36]. Up to 30% of thymoma patients present myasthenia gravis (other patients affected by thymoma can have only AChRAb seropositivity without clinical myasthenia). Thymoma can be revealed by a CT of the mediastinum: normal thymus is usually visible on CT in younger people, but beyond the age of 40, a mediastinal mass in an MG patient is highly suspicious of thymoma. Thymomas originate from the thymic epithelial cells, do not show overt atypia of the epithelial component, typically grow slowly, rarely spreading beyond the thymus, and are correlated to myasthenia gravis and several other autoimmune diseases [37] that can variably associate such as pure red cell aplasia, Good syndrome (thymoma with combined immunodeficiency and hypogammaglobulinemia), Hashimoto's thyroiditis, Isaac's syndrome, POEMS syndrome (polyneuropathy, organomegaly, endocrinopathy, M component, and skin changes) and several others (for a review see [38].

The presence of thymoma is retained to aggravate a prognosis of MG, because of the frequent occurrence of severe symptoms not significantly improved by thymectomy [39].

More recent studies confirm these data [40, 41]. Of note, in thymomas, generally present is an active thymopoiesis with production of CD4 and CD8 T cells having a central role in the pathogenesis of related MG [42-44].

The active thymopoiesis in thymomas would occur because of a defective expression of the autoimmune regulator AIRE [45], a defective generation of regulatory T cells [46] and failure of the positive and negative selection. The spectrum of antibody targets can be very broad in thymoma associated MG in comparison with early onset MG [47] and this could explain the variety of associated autoimmune conditions. For example, striational antibodies that can correlate to expression in thymoma of striational antigens [48], antibodies against ion channels, antibodies against cytokines such as interferon-α, interferon-ω and interleukin-12 [4], antibodies anti glutamic acid decarboxylase or anti collapsin response-mediator protein-5 can be found. Autoreactive T cells from the thymic microenvironment arrive in the periphery, where they can stimulate the pathogenic B cell response [4]. Regional lymph nodes would become the location of the autoimmune process also with the contribution of dysfunctional regulatory T cells and skeletal muscle derived AChR/immune complexes [4]. The process can also continue after thymectomy [4].

Thymic carcinoma is rare, shows histologic atypia, and has a higher incidence of capsular invasion and metastases [49]. It can infrequently develop in thymoma context and is rarely associated to MG [50]. Thymic lymphomas and leukemias originate from T and B lymphoblasts of the thymus and are classified as T or B-lymphoblastic leukemia/lymphoma (WHO 2008) [51, 52]; these malignancies with analogous cases originating outside thymus are rarely found to be associated with MG [53-55].

According to some studies, the pathogenesis of late onset MG without thymoma (LOMG) could be similar to thymoma MG cases [56, 57] in spite of a lack of definite thymic disease. In fact, LOMG patients share with thymomatous MG patients the presence of titin and RyR antibodies (see striational antibodies paragraph for more details), the frequency of an autoimmune response with neutralizing antibodies to cytokines interferon-α and IL-12 [58, 59], similar expansions of peripheral T cell repertoire [4, 60]. It is possible that AChR and titin reactive T cells are generated in an atrophic thymus without adequate expression of AIRE (an age related decline of AIRE positive cells has been described) and become activated in periphery triggering a self-perpetuating mechanism.

The MuSKAb positive MG is different from the AChRAb form having distinct immunologic target and mechanisms for NMJ dysfunction [4]. Thus far, studies do not support a clear role for the thymus in the pathogenesis of MuSK MG in which the risk of thymoma seems very low [4]. Future studies are required to understand mechanisms of pathogenesis in MuSK MG.

In conclusion, MG is an autoimmune disease that can occur as paraneoplastic syndrome thymoma associated. However, the thymus is probably always involved at least in AChRAb seropositive patients.

## 1.4. HUMORAL BIOMARKERS FOR MG

A disease biomarker should be easily detected and quantified with high sensitivity and specificity in serum or other biological fluids. The clinical need for MG biomarkers derives from difficulties in making diagnoses and evaluating a disease course and therapeutic response. In fact, a great clinical variability from one patient to another or in the same patient during the disease course is described. Considering antibody profiles, more specific phenotypes can be delineated (Figure 1). The rare muscle-specific receptor tyrosine kinase antibody (MuSKAb), the seropositive form of MG indeed is almost always generalized with more frequent impairment of bulbar muscles or of neck, shoulder, and respiratory muscles. Required treatment frequently involves immunosuppressive agents whereas acetylcholinesterase inhibitors are of limited utility and thymectomy is generally useless. Plasma exchange seems to be more efficacious than IVIG [61, 62]. Patients with striational antibodies have often a late onset of disease or a thymoma and frequently require immunosuppressive therapy. Patients with Kv1.4 antibodies can be associated with severe MG and heart complications [63, 64]. Patients with low-density lipoprotein receptor-related protein 4 (LRP4) antibodies have often clinical features indistinguishable from that of mild AChRAb + MG whereas cortactin, agrin, rapsyn and collagen Q seropositivities are of unknown clinical significance at present [4, 36, 61, 65] (Table 1).

Thus, specific autoantibodies are markers of specific disease phenotypes. Besides autoantibodies, many other proteins involved in immune response, detected on immune cells or in serum, could have a clinical utility as markers of disease.

Markers of B and T cells are necessary to distinguish between anti- and pro-inflammatory subtypes. In general, an imbalance between regulatory (anti-inflammatory) B and T cells and plasma cells, follicular Th and Th17 cells (pro-inflammatory) is reported in AChRAb MG [66-71]. Increased surface expression of CD21 on AChR specified B cells and decreased surface expression of CD21 on total B cells in generalized MG have been reported [72]. The level of CD21+ AChR specified B cells correlates positively with serum anti-AChR IgG levels. The expression level of CD72 has a negative correlation with anti-AChR antibody levels in MG [73].

As regards immune-related molecules, two recent studies considered the inflammatory circulating protein profile in AChRAb MG. The first [74] reported increased serum levels of a proliferation-inducing ligand (APRIL), cytokines IL-19, IL-20, IL-28A and IL-35. The other [75] showed the increase of matrix metalloproteinase 10 (MMP-10), which is a member of the metalloproteinase family, the transforming growth factor alpha (TGF-α), a growth factor that has important roles for epithelial proliferation and differentiation, and the extracellular newly identified receptor for advanced glycation end-products binding protein (EN-RAGE, also known as protein S100-A12), a protein that binds to calcium, zinc and copper. These last three proteins are involved in the cell cycle progression and differentiation and play multiple roles in the immune response. In the rare MuSKAb MG, scarce data are available but a study [76] demonstrated differences in cytokine patterns, in comparison with AChRAb MG, in experiments of stimulation and immunosuppression of peripheral blood mononuclear cells (PBMC), confirming the distinction between these two forms of disease.

Table 1. Main autoantibody biomarkers of MG

| Marker | Aim | Method | Sensitivity for Generalized MG | Sensitivity for Ocular MG | Sensitivity for Thymoma MG | Specificity for MG in symptomatic patients |
|---|---|---|---|---|---|---|
| AChR Ab IgG1, IgG3 | MG diagnosis | RIPA, ELISA, CBA | 80% (90% and more with CBA) (less with ELISA) [121] | 50% (more with CBA) [109, 104] | <100% [110] | <100% [104] |
| | MG severity: unreliable(PPV 83% NPV 59%) [119] | RIPA, ELISA, CBA | - | - | - | - |
| MuSK Ab IgG4 | MG diagnosis | RIPA, ELISA, CBA | 1-10% (RIPA) [104] 13% (CBA) [171] | -Very rare (RIPA) -27% OMG AChRAb(CBA) [171] | Very rare [173] | <100% [104] |
| | MG severity: unreliable | RIPA, ELISA, CBA | - | - | - | - |
| LRP4 Ab | MG diagnosis | RIPA, ELISA, CBA | 1-5% (more with CBA) (less with ELISA) [104] | Rare [104] | Exceptional [104] | Low (f.p. in MS, ALS, NMO) [104] |
| Agrin Ab | | Immunoprecipitation, ELISA, CBA | 10-15% [265,266] | - | - | High [266] |

| | MG diagnosis | | | | | |
|---|---|---|---|---|---|---|
| Cortactin Ab | | ELISA and western blot | ~11% [284] | 23,5% DSN [284] | - | Low (f.p. in AID and healthy controls) [284] |
| Collagen Q Ab | | CBA | ~3% [274] | - | - | F.p. in epilepsy [274] |
| Titin Ab | | RIPA, ELISA | LOMG 58% [216] | 12.5% of tSN(RIPA, 229) | 95% [216] | High (f.p. in ALS and MS) [229] |
| RyR Ab | | Western blot, ELISA | LOMG 6% [216]-15% [234] | - | 70% [216] | - |
| Kv1.4 Ab | | RIPA | 18% [240] | 20% [240] | -(40-70% in Japanese patients) [63] | -(high specificity in Japanese patients) [63,240] |
| Rapsyn Ab | | Immunoblotting | - | - | 11,7% [217] | Low (positive in 78% SLE) [217] |

Abbreviations: Ab, antibody; AChR, acetylcholine receptor; AID, autoimmune disorder; ALS, amyotrophic lateral sclerosis; CBA, cell based assay; DSN, double seronegative; ELISA, enzyme-linked immunosorbent assay; f.p. false positive; LRP4, low density lipoprotein receptor-related protein 4; MuSK, muscle specific kinase; NMO, neuromyelitis optica; NPV, negative predictive value; PPV, positive predictive value; RIPA, radioimmune precipitation assay; SLE, systemic lupus erythematosus; tSN, triple seronegative.

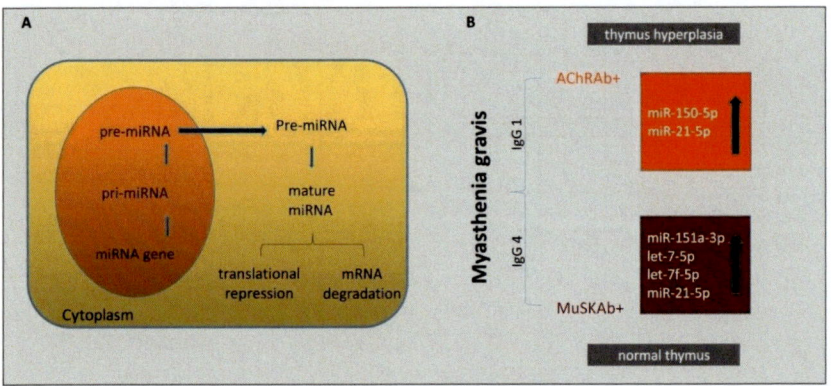

Figure 4. A: Biogenesis of microRNAs. B. Different profile of extracellular miRNAs in serum of MG patients, depending on antibody subtype.

Resistin, a protein involved in an immune response with an apparent pro-inflammatory role, has been advocated to be a potential marker for MG severity and the presence of thymoma [77].

Some authors studied the role of oxidative stress in MG pathogenesis, considering natural antioxidant albumin and uric acid as possible markers [78].

MicroRNAs are small non-coding RNA molecules that realize an epigenetic regulation of gene expression. These molecules are detected in cells and also in the circulation where they are packed into the exosomes to be protected from degradation. Circulating microRNAs (miRNA) have been reported to be biomarkers in several diseases such as cardiovascular diseases, cancer and multiple sclerosis. Some studies have also evaluated their potential as biomarkers in MG. In particular, miR-150-5p and miR-21-5p [79] and the let7 family [80] have been signaled to be increased in sera of AChRAb and MuSKAb MG patients, respectively (Figure 4). These miRNAs are involved in the development of T- and B-cells' autoimmune response. The different profile in AChRAb and MuSKAb MG indicates the etiological difference between these two diseases.

The research on genetic markers in MG has traditionally considered HLA associations: DR3-B8-A1 in early onset AChRAb MG and DR14, DR16 and DQ5 in MuSKAb MG have been consistently reported [4, 81,

82]. Non-HLA markers such as PTPN22, TNIP1, CTLA4 genes and several others have also been signaled and are the object of active research [83, 84].

In summary, potential humoral biomarkers of MG are all the molecules involved in its pathogenesis. The aim of this review is to assess the evidence base for the use of main current and potential biomarkers of MG.

# Chapter 2

# ACETYLCHOLINE RECEPTOR ANTIBODIES

The acetylcholine receptor (AChR) is a $Na^+$ and $Ca^{2+}$ channel, composed of different subunits ("heteropentamer") situated at the top of junctional folds of the postsynaptic membrane of the neuromuscular junction (NMJ). This receptor is named "nicotinic" (nAChR) because it is particularly responsive to nicotine and of "Nm type" to distinguish the nAChR of NMJ from the heteropentameric nAChR (composed by α and β subunits) of neuronal cells and from the ancestral homopentameric nAChR (composed by only α subunits, usually α7) expressed by neuronal and non neuronal cells [85, 86]. The nAChR of NMJ is composed of 5 polypeptide chains of 4 types: 2 α chains and one each of β, γ and δ. In adults, γ subunits of embryos are substituted for by ε subunits. Subunits are arranged in a ring around a narrow membrane pore [86]. Each subunit has four membrane-spanning segments, M1–M4. The second membrane-spanning segment, M2, contours the lumen of the pore, and forms the gate of the closed channel [87]. Alpha subunits have an extracellular cysteine loop that mediates the binding of acetylcholine [88]. When acetylcholine binds to N termini of α subunits, the channel changes from a closed (or resting) state to an open state, cations (mainly $Na^+$) enter into the cell and a depolarization (end-plate potential) occurs; consequently, voltage-gated Na channels open permitting the action potential and the muscular contraction.

The channel opening is very fast (few microseconds) when acetylcholine binds to the receptor, then acetylcholine concentration decreases because of hydrolysis by acetylcholinesterase and diffusion and channel closing follows [89] in relation to the dissociation of acetylcholine-AhR binding.

Acetylcholine receptor antibodies (AChRAb) are the main biomarkers of MG. These antibodies are of the IgG1 and G3 types and bind divalently to AChRs inducing their internalization and so accelerating their turnover [90], or can activate the complement membrane attack complex and macrophage attacks (destruction of junctional folds, increase of distance from pre- to postsynaptic membrane increasing the distance that acetylcholine molecules have to cover, result from this mechanism) [91], or can also block the binding of acetylcholine to AChRs (this type of antibody seems to be rare ) [92]. Three mechanisms of damage and consequently three types of autoantibodies are recognized, modulating, binding and blocking [93]. Damage to the postsynaptic membrane due to complement activation is retained in the major pathogenic mechanism, in fact complement-deficient mice are resistant to EAMG induced by AChR immunization [94, 95].

As regards epitope binding, antibodies against AChR are heterogeneous. The majority binds to α subunit and precisely to the main immunogenic region (MIR) [95]. The MIR consists of a group of overlapping conformation dependent epitopes around the 67–76 amino acids of the extracellular domain of α subunit, but other segments also contribute to this antigenic region [96]. Autoantibodies to other epitopes of α subunit and of the other subunits have been identified in the serum of MG patients [97, 98]. However, the AChRAb against α subunit is the most pathogenic, particularly in comparison with that against β subunit. This seems to derive from the location of MIR on the top of the AChR that favors the cross-linking of AChRs by the anti-MIR antibodies and consequently AChR destruction by both complement and antigenic modulation mechanisms. The AChRAbs against the β subunits are less capable in cross-linking the AChRs and there is therefore less damage to AChR. The AChR epitope pattern influences disease severity [97, 99] and a MIR Ab assay has been proposed to be useful for predicting MG severity [100].

It seems that at the onset of MG the AChRAbs are focalized to single epitopes of the α subunit but successively, the antibody response may also spread to other epitopes [101] because of the secondary involvement of AChRs derived from muscle or thymic myoid cells [4, 102, 103].

The causality link between AChRAbs and MG and the above-mentioned considerations about the pathogenicity of AChRAbs derive from animal studies, *in vivo* and *in vitro*. Both experiments of active immunization against the AChR and passive transfer with rat- or mouse-derived monoclonal AChRAbs, and passive transfer of human purified AChR specific immunoglobulins to animals have all been informative [104-106]. For immunization, both native AChR pentamer and recombinantly expressed subunits of the AChR have been utilized [107].

The reference methodology for determination is radio immune-precipitation assay (RIPA) [108]. This test has a sensitivity of 85% in generalized MG, lower in the ocular form (about 50%) [109] (when positive in ocular MG, the probability that MG will become generalized is increased) and near 100% in MG with thymoma [110] (but isolated cases of AChRAb seronegative myasthenia gravis associated with thymoma have been described) [111]: when positive, is overwhelming any other diagnostic procedure. Specificity is practically 100% in all forms of MG [36], although rare false positivity can be detected in autoimmune liver disease, systemic lupus, rheumatoid arthritis patients receiving penicillamine, in allogeneic bone marrow transplantation patients who develop graft-versus-host disease, in patients with thymoma without MG, and in neuromyelitis optica [112]. False negativity may be related to immunosuppression or an early blood test. In RIPA, the AChR is mixed with 125I-alphabungarotoxin (that is a competitive antagonist to nAChR and binds to a site different from that of antibody) and, after incubation with patient serum, a second antibody is added to precipitate the complex AChR-125I-alphabungarotoxin-antibody; then, the precipitate is counted and compared to healthy controls [36, 113]. Both adult and fetal antigens have to be utilized to detect respective antibodies and increase sensitivity.

Detection of antibodies against both (fetal and adult) AChR isoforms in ocular MG implicates a greater risk of generalization [114]. The above mentioned assay measures binding AChRAbs and is the most commonly utilized and useful [112]. Two other methods have been realized to measure blocking and modulating AChRAbs, respectively determining the inhibition of $^{125}$I-α-bungarotoxin labeling of AChR by patient serum and the quantity of internalized, processed $^{125}$I-α-bungarotoxin-labeled AChR released from cultured cells (but this last assay cannot distinguish blocking and modulating Abs). These assays are useful in cases of negativity of AChR-binding assay adding a relatively small diagnostic contribute [112]. Another method of AChRAb determination is the enzyme-linked immunosorbent assay (ELISA) [115] in which antigen coated wells are filled with the serum to be tested, then a specific antibody linked to an enzyme (alkaline phosphatases or horseradish peroxidase) is applied; after washing, the enzyme's substrate is added, producing a detectable signal (usually a color change). In comparison with RIPA, there are no radioactive isotopes but the cut-off levels of ELISA are difficult to determine [36]. Results of the two tests are not identical, ELISA has more false negatives and also false positives and in clinical practice RIPA is the gold standard. There is not a correlation between serum AChRAb titers and the severity of the MG condition when different patients are compared [116] but fluctuations in AChRAb levels in an individual patient could correlate with the severity of muscle weakness and to predict exacerbations [117]. Generally, patients have higher titers before immunosuppressive treatment than after, during stable remission. For example, an old study found a correlation between excessive rebounds of AChRAb level at the first week after double filtration plasmapheresis and clinical worsening and further increases in AChRAb level at the first month and a poorer outcome [118]. However, more recently [119], changes in AChRAb levels of patients from a prospective trial in MG and from another 85 patients of an MG Clinic were evaluated: AChRAb levels fell in 92% of patients who improved and in 63% who did not, thus the positive predictive value for clinical improvement was 83% and the negative predictive value was only 59%. Consequently, AChRAb levels are not recommended as biomarkers of improvement in MG [119]. The effective loss of functional AChRs is the most important factor correlating to the severity of MG [120].

Patients negative for RIPA can be studied with the cell based immuno fluorescence assay (CBA) method [121]. This assay permits the expression of AChRs on the surface of human embryonic kidney (HEK) cells, doubly transfected with the cDNA encoding this antigen and rapsyn (responsible for the aggregation). The binding of antibodies can then be scored visually with indirect immunofluorescence (by fluorescent secondary anti-human IgG antibodies). A range of 16-60% of negative patients are found to be positive with this method mainly because antigens are presented clustered on the cell surface [122-124]. It is possible that there is a divalent binding between sample antibodies and adjacent AChRs that are closely packed on the cell surface [122]. Clustered AChRAbs are mainly of the complement-fixing IgG1 subtype, as in RIPA positive AChRAb MG patients, and in experiments of passive transfer to mice can determine defects of neuromuscular transmission, that has been confirmed with electro-physiological methods [122, 125]. It seems that RIPA seronegative patients with CBA detection of AChRAbs have relatively mild disease and ocular MG [124]. Other authors analyzed the clinical features of clustered AChRAb MG. These patients can present early onset MG and thymic hyperplasia, late onset MG and thymic involution, or thymoma associated MG. Mild forms of MG are more common with a satisfying response to cholinesterase inhibitors and immunosuppressants [123].

Myasthenic patients seropositive for AChRAb can present additional autoantibody specificities; some patients can develop other autoimmune diseases. A study assessing autoimmune associations in a case series of MG [126] found that 18% of early onset patients have a second autoimmune disease (more commonly, thyroid disorder and systemic lupus erythematosus; two cases of neuromyelitis optica) and 40% have a blood relative with autoimmune disease (more commonly thyroid diseases, rheumatoid arthritis and insulin dependent diabetes mellitus), mainly with the R620W risk allele PTPN22. More frequent autoantibodies were against adrenal cortex, thyroid peroxidase, thyroglobulin, nuclear antigens, mitochondrial antigens. In late onset patients, 13% had another autoimmune disorder and 20% had a blood relative with autoimmune disorder; striational autoantibodies were present in 61% of patients (less frequently anti type I interferons and IL12, anti nuclear antibodies, p- or c-ANCA). In a thymoma group, 87% were affected by MG whereas only 8% had blood relatives with autoimmune conditions (thyroid disease and insulin dependent diabetes mellitus; multiple sclerosis in three cases); striational autoantibodies were present in 62%, anti type I interferons, IL12, thyroid, adrenal cortex antibodies in some cases. Authors found that usually, autoimmunity is sharply focused in MG patients rather than randomly adressed.

The rare contemporaneous presence of AChRAb and antibodies against the water channel aquaporin-4 (AQP4Ab), considered the main etiologic factor for neuromyelitis optica (NMO), has been investigated in depth. The cell based assay is the most sensitive, 91%, and specific, 100%, assay for AQP4Ab [127]. According to some authors, MG and NMO seem to occur together 70 times more frequently than would be expected by chance in the British population [122]. Neuromyelitis optica in MG has been reported mainly after the onset of MG and after thymectomy [128]. In MG patients with symptoms/signs of central nervous system involvement, AQP4Abs are a marker for NMO comorbidity.

In conclusion, AChRAb is a good biomarker for the diagnosis of MG and its presence reinforces the need for excluding a thymoma (that however needs to be excluded also in AChRAb seronegative MG cases). These antibodies are not reliable as a severity index comparing different patients, but can correlate with relapses in the same patient.

# Chapter 3

# MUSCLE-SPECIFIC RECEPTOR TYROSINE KINASE ANTIBODIES

The muscle-specific receptor tyrosine kinase (MuSK) is a 100 kDa transmembrane protein fundamental for the formation and maintenance of NMJ and is situated in its postsynaptic side. The molecule of low-density lipoprotein receptor-related protein 4 (LRP4) forms a homodimer, which combines with a homodimer of MuSK constituting a tetrameric protein complex on the postsynaptic membrane of the NMJ [129]. Experimental studies showed that agrin, a proteoglycan secreted by the motor nerve terminal, binds to LRP4 which consequently causes the activation of the protein tyrosine kinase function of MuSK and its phosphorylation [107]. This latter event causes a progression of the downstream signaling with several molecular consequences among which phosphorilation of DOK7 (a muscle specific adapter protein,) and β2 subunit of AchR [130-134]. Phosphorilation of MuSK is dependent on protein kinase CK2 that interacts and colocalizes with MuSK [135]. Wnt ligands (wnt ligands are molecules that contribute to signal transduction for cell-cell or same cell communication) also directly bind to and phosphorylate MuSK to induce AChR clustering especially at an early stage of development [129]. Then, rapsyn (43 kDa receptor-associated protein of the synapse) is able to form dense postsynaptic clusters of AChRs, linking them to the underlying postsynaptic cytoskeleton, possibly by association with actin or spectrin.

Antibodies against MuSK (MuSKAbs) in MG have been described for the first time in 2001 [136]. These antibodies can be determined by cell based assay or RIPA (using 125I-labeled recombinant human MuSK protein). MuSKAb MG is present in about 1-10% of all MG patients.

These antibodies appear to bind mainly the Ig1 (N-terminal end) but also other domains of MuSK among which is Ig4 (C-terminal end) (there are differences in the epitope specificities of the MuSK-IgG in different reports/patients that may indicate heterogeneity of the anti-MuSK antibodies or other causes such as the assay system [129]), thereby impeding formation of the agrin-LRP4-MuSK complex [107, 129], i.e., the agrin-dependent MuSK activation [137, 138]. Blocking of the postsynaptic tyrosine phosphorylation causes accelerated loss of AChRs from the postsynaptic AChR cluster [139, 140], up to the failure of neuromuscular transmission [141]. Moreover, the blocking of MuSK-ColQ interaction has also been reported as a consequence of MuSKAbs by an *in vitro* binding assay [142], but subsequent experiments have demonstrated that this has no essential effect on AChR clustering [129].

These effects of MuSKAbs have been studied in MG patients and animal models. Indeed, experiments in mice, rats and rabbits demonstrated that an active immunization with MuSK protein can induce NMJ impairment with a decrease of AChRs and EPPs [143-148]. Ultrastructural and functional studies have indicated both pre and postsynaptic defects [147]. Passive transfer of Ig in mice can induce the same consequences [139, 141, 142, 149-152].

However, AChR deficiency has not been confirmed by biopsies of intercostal muscle [153] or of biceps brachii muscle [129, 154] in MuSK-MG patients. This apparent contradiction can be related to different MuSK expressions in analyzed muscles [155]: MuSK expression is high in the soleus, medium in the intercostal muscle, and low in the omohyoid muscle [155]. Selective muscle distribution of deficits in MuSKAb MG (and in some cases of congenital myasthenic syndrome with mutations of MuSK) could be explained by different endogenous levels of MuSK that could correlate with the muscle-specific differences in the response to increased agrin-MuSK signaling [155].

In theory, the hindrance of ColQ-MuSK interaction by MuSKAbs should reduce the expression of AChE at the NMJ and cause lack of the effects of cholinesterase inhibitors in MuSK-MG patients. Nevertheless, deficiency of AChE observed in the mice model has not been proven in MuSK-MG patients.

It is recognized that antibodies in MuSKAb MG are mostly IgG4 although also IgG1 and IgG3 may be present [136, 156]. These Ig subclasses have different Fc regions [157]. Immunoglobulin G1 and 3 are bivalent, activate complement and immune cells. Immunoglobulin G4 is monovalent due to Fab (fragment, antigen binding) arm exchange [158], cannot activate complement and has low affinity for Fc receptors on immune cells, and is considered anti-inflammatory [159].

A study [160] claimed that MuSK immunoglobulin-G4, but not immunoglobulin-G1-3, binds to mouse neuromuscular junctions *in vitro*, and that injection into immunodeficient mice causes paralysis and reduction of postsynaptic AChRs density; electrophysiological analyses revealed pre and postsynaptic defects and impairment of compensatory mechanisms (compensatory transmitter release upregulation, present in AChRAb MG).

Another study [161] showed that IgG4 blocked binding of LRP4 to MuSK but also that IgG1 and 3 could impair clustering of AchRs. Moreover, both the IgG4 and IgG1-3 fractions were demonstrated to cause dispersal of the AChR clusters in a mouse muscle cell line by mechanisms likely independent of the interaction with LRP4. Perhaps IgG1-3 can bind MuSK divalently, and activate phosphorylation, causing desensitization of MuSK with a loss of function, is only a hypothesis [2].

A very recent study that realized an experimental autoimmune MG in IgG1 deficient mice (IgG1 in mouse is the analog of IgG4 in humans), confirms that also complement fixing antibodies can cause MG. In this model, the predominant type of antibody is IgG3 activating complement through both the classical and alternative pathways. The study shows also that levels of MuSK-reactive B cells correlate better than MuSKAbs with the degree of MG severity suggesting that MuSK-binding B cell measurements might be used as a potential prognostic biomarker for MuSK antibody positive MG. Autoreactive B cells have many pathogenic features: they produce cytokines and can present antigens. However, complement fixing antibodies can cause MuSKAb MG and their role must be elucidated [162].

Moreover, MuSKAbs have been demonstrated to upregulate muscle RING finger protein 1 (MuRF-1) in muscular cells and atrogin expression in TE671 cells *in vitro* (a human cell line derived from a medulloblastoma), conducing to muscle atrophy [163-165].

Epidemiological studies on MuSKAb MG report very variable data on prevalence: initially, it was reported in 70% of AChRAb negative MG patients [136] but subsequent data were always inferior with very variable percentages among ethnique/racial groups: South Europeans are more frequently positive (in Italy 47.4% see reference 166) than Northern Caucasians (only three patients, two Caucasians, reported in the well-defined Norwegian population of 5 million inhabitants [167]). MuSKAbs were present only in 8.7% of a cohort of 48 AChRAb seronegative generalized MG patients [168]. It seems that in Europe MuSKAb MG is more frequent within latitudes 30° and 50° N of the equator [169]. The frequency of the disease is greater in Afro-American. A frequency of 3.8% was reported in a Chinese cohort [170]. A great screening study using a cell based assay detected MuSKAb in 83/633 AChRAb, MuSKAb and LRP4Ab RIPA seronegative MG patients from 13 countries (13.1%: from 4.8% in the Spanish cohort to 22.4% in the Serbian cohort); moreover, 27% of ocular seronegative patients were MuSKAb positive (pure ocular MG is exceptionally seropositive for MuSKAbs with RIPA), 23% had thymic hyperplasia. Additional subjects positive for 2 or also 3 antibodies were found; some cases of MuSKAb positivity were found in multiple sclerosis or neuromyelitis optica patients [171].

Another study has demonstrated a CBA positivity for MuSKAbs in 63/64 (98%) of RIPA positive MuSKAb MG patients sera and in 11/145 (8%) of MG sera negative for AChR and MuSK Abs by RIPA or AChRAb CBA [125]. The specificity of the assay for MG in these patients was 97%.

The majority of these patients (92%) had MuSK IgG4 abs but many also had MuSK IgG1-3 as has previously been reported [138].

Clinical features of MuSK myasthenia present several peculiar features that have been studied by several research groups and recently in two large cohorts from Duke University and from the Catholic University of Rome.

MuSK antibodies (MuSKAb) have been found in 39% at Duke and 49% at Catholic University in AChRAb negative generalized MG patients with a large prevalence of women [172]. The age of onset of MG in females was distributed around a peak in the fourth decade with a small peak in the second. In this study and in literature, MuSKAb MG is characterized by frequent impairment of bulbar muscles and respiratory crises; frequently atrophy of the tongue is found. Another notable clinical picture is characterized by involvement of neck, shoulder, and respiratory muscles; also cases indistinguishable from generalized AChRAb MG are possible [172].

In literature, the presence of MuSKAbs is found very rarely in conjunction with AChRAbs or with thymoma [173]. Some cases of double seropositivity with AChRAb and MuSKAb were due to antibodies to alkaline phosphatase [174].

The course of MuSKAb MG is frequently characterized by acute onset and rapid progression in the first year followed by stabilization. In some cases, MG is mild/moderate for years and then can aggravate despite therapy. Long-term outcomes were commonly favorable: 54% achieved MGFA-PIS (post intervention status) of minimal manifestations or better.

As regards therapy [172], patients with MuSKAb MG treated with acetylcholinesterase inhibitors have a clinical improvement only in 57% of cases and frequently manifest adverse effects: worsening symptoms (5%), skeletal muscle cramps (9%) and fasciculations (16%). Moreover, patients often necessitate corticosteroids (92% in the above-mentioned study) and immunosuppressive treatments. Response to rituximab therapy, offered to patients with refractory MG, was satisfying without significant adverse effects.

Studies have revealed that MuSKAbs levels could be correlated with MG severity and with response to immunotherapy [172, 175]. However,

MuSKAb absolute levels do not correlate with disease severity [176]. MuSKAb levels fall with treatment, are highly variable and do not correlate well with disease severity or clinical response [177].

In conclusion, MuSKAbs are a biomarker for a well-defined form of MG different from the AChRAb form. These antibodies may tend to correlate with the activity of disease [178].

*Chapter 4*

# LOW-DENSITY LIPOPROTEIN RECEPTOR-RELATED PROTEIN 4 ANTIBODIES

The low-density lipoprotein receptor-related protein 4 (LRP4) is a transmembrane protein member of the Lipoprotein receptor-related protein family, located on postsynaptic side of NMJ where it is specifically expressed by subsynaptic myonuclei.

The protein LRP4 contains a big extracellular N-terminal region with multiple EGF repeats, LDLR repeats, a transmembrane domain, and a short C-terminal region [179].

In addition to NMJ, this protein is found in multiple tissues in mice and has important functions during the development of limbs, lung and kidney [180-183].

In the development and maintenance of NMJ, LRP4 is the receptor for agrin that is secreted by the terminal nerve, and for MuSK, situated in postsynaptic membrane. The molecule of LRP4 forms a homodimer, which combines with a homodimer of MuSK constituting a tetrameric protein complex on the postsynaptic membrane of the NMJ [129]. Agrin binds to the N-terminal region of LRP4 including a subset of the LDLa repeats and the first of four β-propeller domains [184] and increases affinity of LRP4 for MuSK with a strict association of these two molecules (in this step Ig1 and Ig4 domains of MuSK are very important; the domains of LRP4 that bind MuSK are the 4th and 5th LDLa repeats close to the N-terminal end, and the third β-propeller domain) with phosphorylation of MuSK, and the downstream signaling cascade necessary for AChR clustering (see also MuSKab paragraph).

In addition to its postsynaptic roles, LRP4 also has very important presynaptic effects, namely signals to motor axons regulating the development, stabilization and plasticity of NMJs [185]. Synapses between motor axon and muscle happen in a prepatterned postsynaptic region, characterized by AChRs, MuSK, LRP4 and other particular postsynaptic proteins. This postsynaptic "preparation" for definite NMJ depends upon MuSK and Lrp4 but Lrp4 is sufficient to presynaptic differentiation with presynaptic clustering of synapsin, a protein associated with synaptic vesicles, and of active zone proteins.

Considering experimental data as a whole, we should consider the Lrp4 function as very important in three steps during NMJ formation: before innervation, Lrp4 forms a complex with MuSK to realize muscle prepatterning; when motor axons approach muscle, clustered LRP4 (consequence of MuSK activation) acts as a retrograde signal to presynaptic differentiation; when motor axons establish contact with muscle, Lrp4 binds agrin, secreted by nerve terminals, stimulating further MuSK phosphorylation and stabilizing NMJ [185].

Considering LRP4 functions, the autoantibodies targeting it should be pathogenic and this seems to have been proven by many researchers.

Models of experimental autoimmune MG (EAMG) have been realized by actively immunizing mice with ecto-LRP4 [186]. These mice developed clinical and electrophysiological deficits with a structural defect in NMJ. Their antibodies against LRP4 impaired the neuromuscular transmission by interfering with agrin/MuSK signaling and fixing complements. Very recently [187], complement-related damage has been proved again with a demonstration of complement fixing LRP4Abs in sera and complement/ IgG deposits at the NMJ in the actively immunized mouse strain C57BL/6 (B6); clinical features, cytokine patterns and NMJ aspects were very similar to AChRAb immunized mice. Furthermore, if IgGs of immunized rabbits are injected into healthy mice, a passive EAMG model is realized with all the MG-like symptoms and signs [186]. These results convincingly demonstrated the pathogenicity of LRP4 antibodies.

In epidemiological studies, LRP4Abs have been found in 1-5% of all MG patients but in 2-50% of AChRAb and MuSKAb seronegative MG patients with female preponderance [188].

Higuchi et al. [189], for the first time described autoantibodies to LRP4 in 9 patients (3%) in a cohort of 300 AChRAb seronegative patients (Japan); 3 patients out of 9 were also MuSKAb positive; in a cohort of 101 Lambert Eaton myasthenic syndrome (LEMS) patients, one was also weakly positive for LRP4Abs; healthy controls and AChRAb positive MG patients were negative for LRP4Ab. These antibodies were mainly IgG1 directed to the extracellular portion of LRP4 responsible for interaction with agrin. A luciferase-reporter immunoprecipitation (LUCIP) method was used, based on the strong luminescence of Gaussia luciferase (GL) bound to the extracellular portion of LRP4. If present, serum LRP4abs could bind to LRP4-GL and be measured by luciferase activity with specific assay kit to calculate the amount of LRP4-GL protein and infer the titer of LRP4Abs. A contemporary study demonstrated that the extracellular portion of LRP4 is sufficient to regulate both the presynaptic and postsynaptic differentiation in the NMJ [190].

Antibodies against LRP4 can also be detected by a cell based assay or a RIPA [36]. Pevzner [191] used a cell based assay to ascertain LRP4Ab in 6 (50%) out of 13 AChRAb and MuSKAb seronegative MG patients (Germany), in none of 36 AChRAb MG patients, in one MuSKAb MG patient. The different estimates of frequency of LRP4Ab in Higuchi and Pevzner studies were attributed to the different populations considered (Asiatic and Caucasian) and/or the screening method used. Zhang et al. found LRP4Ab in 11 (9.2%) out of 120 AChRAb and MuSKAb seronegative MG patients and in one of 36 MuSKAb MG patients (the Hellenic Pasteur Institute, Greece, and Wayne State University, USA); interestingly, two patients with neuromyelitis optica were also LRP4Ab+ [192].

Tzartos et al. [193] reported LRP4Abs in 24/104 (23.1%) amyotrofic lateral sclerosis patients from Greece and Italy and 4/84 (4.8%) multiple sclerosis patients, using a cell based assay (RIPA method on same population found 11.5% seropositivity in ALS patients; 4 ALS patients were RIPA positive but CBA negative). Seven ALS patients also performed a search of LRP4Abs in CSF and 6 were positive. Patients affected by ALS with or without LRP4Ab were clinically indistinguishable. Authors concluded that LRP4Abs involved in neurological diseases affecting LRP4-containing tissues are more frequent in ALS patients than MG patients and perhaps have a pathogenic activity in ALS in the denervation process. In a later study the authors found 14.9% of LRP4Ab in ALS patients from Israel [194]. Rivner et al. [195] confirmed the presence of LRP4Ab in ALS at a lower frequency (9.8%), explicable with a lesser sensibility of the utilized method (ELISA). The authors hypothesized a pathogenic activity and a role as biomarker for a subgroup of ALS patients.

Some cases with ALS and LRP4Abs that had myasthenic symptoms and a partial response to immunotherapy [196] were also reported.

Other authors using a cell based assay performed a screening of about 800 MG patients from 10 countries. The frequency of LRP4Ab (IgG1 and IgG2 subtypes) in 635 MG patients seronegatives for AChRAb and MuSKAb was 18.7% (range from 7% Norway to 32.7% Poland). The prevalence was higher in women (F/M ratio 2.5/1), age onset 33.4 for females and 41.9 for males. Moreover, 8/107 anti-AChR positive and 10/67 anti-MuSK positive sera also were seropositives for LRP4; also 4/110 from a group with other neuroimmune diseases were LRP4+, all affected by multiple sclerosis. The clinical picture of LRP4 patients was a mild form of MG, similar to a mild form of AChRAb positive MG also for a response to cholinesterase inhibitors or corticosteroids (even if 9.6% of MG patients with anti-LRP4 antibodies did not respond to cholinesterase inhibitors). A percentage of 6.6% of MG patients with anti-LRP4 antibodies demonstrated myasthenic crisis. Patients with AChRAb+ or MuSKAb+ in addition to LRP4 had a more severe form of disease. In 31% of explored thymi hyperplasia was found without thymoma cases [197]. Considering the above-mentioned study with the others in literature, LRP4Ab patients have an ocular or generalized mild form of myasthenia gravis, and about 20% of patients have only ocular impairment for more than 2 years. Respiratory insufficiency is very rare (a case of a 72-year-old man demonstrating dropped head and acute respiratory insufficiency with isolated LRP4Abs has been reported, with good response to steroid pulse therapy, see reference [198]), with the exception of patients with additional MUSK antibodies. In two-thirds of LRP4Ab positive patients, the thymus is normal for the age, but hyperplasia has been reported [188].

In conclusion, LRP4Ab is a biomarker of a form of MG similar to a classically mild form of AChRAb MG and also for their response to treatments. The contemporary presence of MuSKAbs or AChRAbs seems to characterize a form similar to MuSKAb or AChRAb MG, respectively. This antibody can be detected also in ALS and more rarely in multiple sclerosis with an unknown significance if not associated with myasthenic syndrome.

# Chapter 5

# STRIATIONAL ANTIBODIES

## 5.1. INTRODUCTION

Striational antibodies (StrAbs) target intracellular epitopes on skeletal muscle proteins including titin, myosin, tropomyosin, actin, α-actinin, actomyosin, rapsyn and ryanodine receptors [199-201], not directly accessible to autoantibodies. Alpha subunit of membrane voltage-gated K channel, Kv1.4, has been added to this group.

This group of antibodies are found in 30% of MG patients and in 90% of thymoma MG (T-MG) [202], generally in the late onset group with equal gender ratio, and infrequently in other paraneoplastic/autoimmune neurological disorders (e.g., Lambert-Eaton myasthenic syndrome) and also in thymoma without MG. A study investigated the positive (PPV) and negative (NPV) predictive value of AChRAbs and StrAbs for thymoma presence in MG: PPV of binding AChRAbs plus StrAbs was 50% if MG onset <40 years; PPV of all antibodies was <9% if MG onset >40 years; NPV of binding AChRAbs was 99.7% for all ages; modulating and blocking AChRAbs do not increase PPV or NPV for thymoma in all ages [203]. Another study [204] investigated associations of StrAbs in a paraneoplastic serological evaluation. The authors found a significant classical association with MG and thymoma; interestingly a relative frequency of immune disorders (rheumatoid arthritis, pernicious anemia, hypothyroidism) and scarcity of cancer diagnoses in patients with StrAb titer <1:7680 without coexisting paraneoplastic antibodies were reported. Striational antibodies can associate in the same patient: in a study, 8 out of 209 Japanese MG patients resulted seropositive for AChRAb, anti-titin, and anti-Kv1.4 antibodies [205]. Multiple StrAbs are typical of T-MG [40, 206]. According to some researchers, StrAbs could contribute to severe MG symptoms, autoimmune myositis and/or cardiomyositis [207-209]. In fact, inflammatory myopathies can rarely associate with StrAb MG; in these cases, biopsy of skeletal muscles showed CD8+ lymphocyte infiltration. Skeletal muscle cells of MG patients can express MHC class I, and less frequently class II, associated with lymphocytic infiltrations and often thymoma. This common phenomenon implies that muscle cells could act as antigen presenting cells in MG and thus could have an important role in the pathogenesis of the disease [210]. However, it must be considered that some authors signaled that lymphocytic infiltrates of thymoma patients with or without MG are naïve CD45RA+ lymphocites derived from lymphocyte-rich thymoma tissue and are different from true polymyositis [211].

However, there is not conclusive evidence about the real clinical pathogenic effects of striational antibodies [212] although *in vitro* complement activation and T cell proliferative response to MGT-30 (the immunogenic peptide of titin) have been documented [206, 213, 214].

A brief reference of rapsyn antibodies will be made because of the role of this protein in the clustering process of AChRs but at present its antibodies do not seem to have clinical importance in the evaluation of MG. Literature on striational antibodies mainly regards titin, ryanodine receptor and Kv1.4 antibodies.

## 5.2. RAPSYN ANTIBODIES

The receptor-associated protein of the synapse (rapsyn) is a postsynaptic membrane protein of 43 kDa. This protein associates with AChRs and is necessary for their clustering. Its RING-H2 domain binds to dystroglycan, linking rapsyn to the subsynaptic cytoskeleton [215]. Antibodies against rapsyn are found in almost 15% of MG patients [216], most commonly in thymomatous MG but can be associated with systemic lupus [217]. At present, rapsyn antibodies have no role in the clinical evaluation of MG.

## 5.3. TITIN ANTIBODIES

Titin is the largest known protein (isoforms vary in molecular weight between 3000-4200 kDa), spanning from the Z-disc to the M-line in the sarcomere. It is produced in several isoforms with different molecular weight in skeletal and cardiac muscle, as a result of alternative splicing. It is essential for muscle contractility.

Titin has several epitopes but the main immunogenic region (MIR) is a 30 kDa peptide called myasthenia gravis titin-30 (MGT-30) and is situated near the A/I-band junction [206, 218, 219]. Autoantibodies to titin are now determined by an enzyme-linked immunosorbent assay (ELISA), in which a recombinant MGT30 protein is used as the antigen. These antibodies have been shown to be IgG1, and consequently could activate complement. In a study, electromyographic evidence of myopathy in MG patients with titin Abs has been reported [220]. Antititin Abs were present in 27% of a series of 295 MG patients and in 54% of the subgroup with thymoma [221]. These antibodies are a particularly useful biomarker in patients with thymoma as their presence in early onset MG is 99% specific for thymoma; in late onset MG specificity falls to 75% [36]. Epitopes of titin have been identified in thymoma tissue [222, 223]. It is generally admitted that the thymoma pathogenic microenvironment induces aberrant immunization of T-cell against autoantigens [40]. Striational auto-antibodies to titin are also produced by plasma cells derived from thymic B cells of patients with T-MG [224]. In clinical practice, the presence of StrAbs and computed tomographic scans of the anterior mediastinum have a similar sensitivity for revealing a thymoma in MG patients. Very late onset MG patients (>60 years) have anti-titin antibodies in 60-80% of cases [221, 225]. Anti-titin antibody is associated with HLA DR7 in Caucasian MG patients [225-227], with DQA1*02 allele in a Turkish cohort of MG patients [228] and perhaps with DR2 in Japanese [212]. The presence of titin antibodies has been thought to be related to more severe symptoms but this relation is not confirmed in all the studies [221]. In a recent study, radioimmunoprecipitation assay (RIPA) using 125I-labeled MGT30 has been utilized for screening of 667 MG patient sera from 13 countries and titin antibodies not only in AChR+ MG patients (40,9%), but in several seronegative (50/372), as well as in MuSK+ (15,4%) and LRP4+ patients (16,4%) were detected [229]. This method revealed great sensitivity: several cases of titin seropositivity in AChRAb seronegative patients were found because of a lower threshold of detection in comparison with ELISA (ELISA confirmed only the cases with higher titers). The test also has a remarkable specificity: none of healthy controls and myopathy patients resulted as seropositive. Important conclusions of this study were that titin Abs that were positive equally in EOMG and LOMG, do not seem to be a marker of thymoma or severe MG (perhaps in these cases presence of both AChRAbs and titin Abs is required) and that ~13% of patients seronegative for AChRAb, MuSKAb and LRP4Ab are titin antibody positive.

## 5.4. RYANODINE RECEPTOR AUTOANTIBODIES

A ryanodine receptor (RyR) is an intracellular calcium channel (mediates the release of calcium ions from the sarcoplasmic reticulum into the cytoplasm in the process of muscle contraction) of which there are three major isoforms: RyR1 of skeletal muscle, RyR2 of myocardium and RyR3 expressed more widely but especially in the brain. The RyR contains 5035 amino acids, has a molecular weight of 565 kD and is composed by 4 homologous subunits forming a channel [206]. Main immunogenic regions are the peptide chain 2, RyR type 1 fusion protein, and the peptide chain 25 [206]. Antibodies to RyR are IgG1 and IgG3 and could activate complement, provoke allosteric inhibition of RyR, blocking calcium release. Impaired excitation-contraction (E-C) coupling in addition to neuromuscular transmission failure has been reported with RyRAbs [230, 231]. Autoantibodies against dihydropyridine receptor, which have functional interactions with RyR1 in Ca2+ release, were also detected in MG patients (37% of thymoma MG patients, all positive also for AChRAb and RyR antibodies [232, 233].

Antibodies against RyR are detected in 13–38% of all MG patients, associated with MG with late onset (14% of late onset MG patients), thymomas (70% of thymoma-MG patients comprising also invasive/malignant type) and frequently severe MG with bulbar, respiratory and neck impairment [234]. The combination of titin and RyR antibody positivity in MG is 95% sensitive and 70% specific for thymoma [112]. Epitopes of RyR have been identified in thymoma tissue [235, 236]. The same antibodies to RyR1 of skeletal muscle can react to RyR2 of myocardium (cross reaction) in spite of antigenic differences [237]. ELISA based on known epitopes is the laboratory method of choice for detection; another possibility is western blot.

## 5.5. Kv1.4 Autoantibodies

Voltage-gated K channel (VGKC) consists of four transmembrane α-subunits that combine as homo- or heterotetramers. Kv1.4 is a α-subunit with a molecular weight of 73 kD located mainly in the brain (axonal membranes or near axons), peripheral nerves, and skeletal and heart muscles [238]. A RIPA method using rhabdomyosarcoma cell line RD and leukemia cell line K562 as antigen sources can be used to detect anti Kv1.4 antibodies [36].

Anti-Kv1.4 antibodies are detected in 12–15% of MG patients [239, 240] also in association with thymoma and the presence of Kv1.4 mRNA has been demonstrated in thymoma tissue [241]. Anti-Kv1.4-positive patients show frequencies of bulbar involvement and myasthenic crises of 73% and 31%, respectively [206]. Some MG patients had very dangerous arrhythmias such as ventricular tachycardia, sick sinus syndrome, and complete atrial ventricular block [205, 241-244].

In Japan, a study on Japanese patients [64] supported that KV1.4 antibodies were associated with severe MG and heart diseases; at another study Kv1.4 antibodies were also associated with more severe MG than anti-titin antibodies [205]. These data were not confirmed in a study on a European population [240].

In a case report study, serial measurements of cardiac troponin I in MG-related cardiomyopathy were shown to be useful in monitoring for cardiac damage during MG-related therapy [245].

One must also consider the rare but documented possibility of a false positive result of cardiac troponin T due perhaps to interference of heterophilic antibodies, or other antibodies related to MG [246].

## CONCLUSION

The presence of StrAbs may associate with more severe disease [206, 237, 247, 248] sometimes including myopathy and/or cardiopathy.

Myasthenic patients with these antibodies usually require immunosuppressive treatments. The preferred therapeutic scheme is a combination of prednisone with other immunosuppressive agents maintaining low-medium dosages. Tacrolimus, a calcineurin inhibitor (CNI), acts as an enhancer of RyR-related Ca2+ release from the sarcoplasmic reticulum and is widely used [249]. A retrospective study [250] analyzed the 6-month effect of calcineurin inhibitors (tacrolimus and cyclosporine) in 62 StrAbs MG patients. Calcineurin inhibitors were safe and effective although poor responders were 35.5%. Early stages of MG, thymoma and seropositivity for anti-Kv1.4 antibodies were associated with response to CNIs.

The long-term prognosis of StrAbs MG patients is so far not clear. Side effects of immunotherapies may be important factors in morbidity and mortality associated with MG. However, some late-onset MG patients with anti-titin antibodies have a pure ocular form for many years.

*Chapter 6*

# AGRIN ANTIBODIES

Agrin is a large heparan sulfate proteoglycan (three potential heparan sulfate attachment sites in the primary structure of agrin, two of which are generally occupied) involved in the development and maintenance of NMJ in embryogenesis and throughout the course of life. It has a molecular weight of 400 kDa and is composed of a number of structural modules such as other basal lamina proteins. The N-terminus of agrin has alternative start sites generating two isoforms of different length and tissue specificity. The short isoform presents an N-terminal transmembrane region and is chiefly expressed in the brain, whereas the long isoform, which does not have the transmembrane region, is expressed at NMJs and in other tissues [251].

The C-terminus contains three conserved alternative splicing sites, termed X, Y, and Z in mammals. An eight amino acid insert at the Z site lies within the C-terminal LG3 domain and is absolutely required for agrin's AChR clustering activity [252-255].

Several studies suggest a critical role of the agrin/LRP4/MuSK pathway in the formation, maintenance and regeneration of the NMJ (for more details, see paragraphs on MuSKAb and LRP4Ab). The ablation of genes for agrin or for LRP4 or MuSK prevents the formation of NMJ [256-262]. Consequently, agrin is one of the potential target antigens in MG.

Anti agrin antibodies have been demonstrated to inhibit MuSK phosphorylation and AChR clustering *in vitro* [263-264]. Classification of IgG subtype(s) of antibodies against agrin has not yet been explored [4].

A preliminary report utilized a cell based assay fusing the secreted form of neural agrin containing the exons important for clustering the AChR, with the carboxyl-terminal Caspr2 (contactin-associated protein 2): so, agrin is expressed on the surface of HEK293 cells transfected with this molecule. The presence of anti-agrin antibodies in 15% of MG patients was demonstrated (in some cases in association with AChRAb) but sufficient clinical details were lacking [265].

Gasperi et al. searched antibodies against agrin on the basis of its role in NMJ formation and maintenance [266]. They tested 54 MG patients, 30 of which were AChRAb and MuSKAb seronegative, 9 AChRAb positive, and 15 MuSKAb seropositive. One additional patient affected by Pompe disease and 16 healthy volunteers served as control. The authors detected with ELISA elevated titers of agrin Abs in 4 MuSKAb positive patients and one with AChRAb, all of which with a generalized form of MG. Two sera were positive for LRP4Abs with a patient who presented triple seropositivity for agrin, AChR and LRP4 antibodies. One patient (early onset MG) had a thymo-follicular hyperplasia, three had a late onset form. Some agrin antibody positive sera were able to label adult mice NMJs and react with mini agrin expressed by transfected 293HEK cells. One patient who was ELISA negative showed agrin Abs positivity to transfected 293HEK cells and labeled adult mouse NMJs.

Zhang et al. using ELISA confirmed by immunoprecipitation studied 93 serologicallly characterized MG patients [267]. The authors detected seropositivity for agrin antibodies in 7/93 (7-8%) MG patients: 2/4 of AChRAb, MuSKAb, LRP4Ab seronegative, 5/83 AChRAb positive but none of 6 MuSKAb. In this study, there were no LRP4Ab positive patients. The agrin autoantibodies could recognize agrin in its native form on transfected HEK293 cells. These antibodies inhibited agrin-induced MuSK phosphorylation and AChR clustering in cultured myotubes supporting a pathogenic role in MG.

Antibodies against agrin have been reported in ALS patients and thus do not seem to be specific for myasthenia. Generally, from results of literature, agrin antibody positivity associates with other antibodies against proteins of NMJ.

In conclusion, the role of agrin antibodies as a biomarker for a particular form of MG is not until now confirmed.

# Chapter 7

# COLLAGEN Q ANTIBODIES

Collagen Q (ColQ) is composed of three identical subunits containing proline-rich attachment domains (PRADs) binding three tetramers of acetylcholinesterase (AChE). This molecule is present only in the extracellular matrix of NMJ, accessible to antibodies, and concentrates and anchors the enzyme acetylcholinesterase with its catalytic subunit. Collagen Q is bound to NMJ by two heparan sulfate proteoglycan-binding domains (HSPBDs) interacting with the heparan sulphate proteoglycan–perlecan in the basal lamina [268, 269], and by the C-terminal domain (CTD) interacting with MuSK [270]. Both the anchoring systems are necessary to the AChE/ColQ complex; for example, mutations in CTD compromise the anchoring of AChE/ColQ complex to NMJ [271]. Mutations in COLQ gene cause a form of congenital myasthenia because of a deficit of AChE in NMJ [272]. Equally, antibodies to collagen Q may be pathogens because of the decreased concentration of AChE [273, 274]. The MuSKAbs also can disrupt Collagen Q-Musk interaction causing AChE deficiency at the NMJ according to *in vitro* studies [275] and experiments of passive transfer or active immunization for MuSK in mice [276]. In theory, impairment of the AChE function could also derive from antibodies directed to end-plate acetylcholinesterase that effectively have been reported in MG patients, but their presence has been signaled in other autoimmune diseases and healthy controls [277].

The deficit of AChE could provoke a prolonged lifetime of acetylcholine (ACh) in NMJ, an increased duration of endplate currents and an increased ion flux, resulting in compromised neuromuscular transmission or desensitization of the endplate receptors [274, 278].

Recently, collagen Q antibodies have been detected in MG patients and controls with a cell based assay [274]. Human embryonic kidney cells (HEK293) were cultured and transfected with a construct expressing COLQ fused with the transmembrane domain of contactin-associated protein-like 2 (CASPR2).

The screening was applied to 415 MG sera for the presence of antibodies to COLQ. Twelve out of 415 samples were found to be seropositives for ColQ antibodies: 5 AChRab- MuSKAb-, 5 also AChR-Ab+, 2 also MuSKAb+. Considering only the 149 AChRAb- MuSKAb-LRP4Ab-agrinAb- patients, the percentage of ColQAb positivity is 3.4%. The control group was composed of 22 healthy subjects and 21 epilepsy patients, one of which resulted mildly positive.

All of the positive patients were female, and in 8 (out of the ten with known age of onset) MG onset was below 40 years. Three had ocular MG, 7 generalized MG, or 2 ocular/bulbar MG. The patients were treated with steroids or AChE inhibitors: 7 patients responded well to steroids and 4/9 patients, with sufficient information, had a poor response to AChE inhibitors. This reaction to AChE inhibitors could be interpreted on the basis of pathogenic mechanisms of ColQ antibodies.

At present, there are not sufficient data for considering antibodies against ColQ a biomarker for MG; additional evidence is necessary to clarify their real value in clinical practice.

*Chapter 8*

# CORTACTIN ANTIBODIES

Cortactin is a monomeric protein present in the cytoplasm of all cell types. Its structure comprises an amino-terminal acidic (NTA) region; several 37-residue-long repeats; a proline-rich region; and an SH3 domain. It is activated by phosphorylation, by tyrosine kinases or other kinases, in response to extracellular signals. When activated, cortactin can promote polymerization and rearrangement of the actin cytoskeleton recruiting Arp2/3 complex proteins to existing actin microfilaments, facilitating and stabilizing nucleation sites for actin branching. Cortactin is involved both in the formation of NMJ and the clustering of postsynaptic AChRs. Cortactin mediates the assembly of the cytoskeleton triggered by the synaptogenic signal that is dependent on phosphorylation, downstream from agrin/MuSK until AChR/rapsyn clustering [279]. Agrin, added to myotubes *in vitro*, enhances the phosphorylation of cortactin. A mutant form of cortactin (phosphorylation defective) in muscle cells strongly inhibits AChR clustering in response to innervation [280].

Gallardo et al. [281], using a protein array approach found cortactin as a candidate antigen in seronegative MG. The authors revealed with ELISA (confirmed by immunoblot) the presence of cortactin antibodies (cortAbs) in 18/91 (19.7%) AChRAb- and MuSKAb- MG patients, 5/69 AChRAb+ MG patients and 0/34 MuSKAb+ patients; in addition, rare cases of

Lambert-Eaton myasthenic syndrome and autoimmune disease were found to be positive for cortactin antibodies. One out of 19 (5.2%) healthy controls was positive, too. No patient was LRP4Ab+. The authors concluded that perhaps cortactin is implicated in the pathogenesis of MG but could also be a consequence of epitope spreading. The presence of this antibody in some cases of immune diseases could indicate a particular autoimmune background in some patients. Although a distinctive MG phenotype is not evident in this study, cortactin antibodies could be a marker of autoimmune disease and their presence could support the use of immunomodulatory treatment.

At the same time, cortactin antibodies were found by ELISA and immunoblot in 20% (7/34) of patients with polymyositis but also in some patients with dermatomyositis and immune-mediated necrotizing myopathy. Rare cases were evidenced in patients with autoimmune diseases [282].

More recently [283], the same authors of the first report about cortactin showed cortactin antibodies with ELISA and western blot in 9/38 patients with AChRAb- MuSKAb- MG (23.7%; in particular, 4/17 double seronegative patients with ocular MG were cortactin antibody positive, that is 23.5%), with a female prevalence. These patients were younger than AChRAb+ patients, had an ocular or mild generalized phenotype of myasthenia gravis (I or IIA) without bulbar signs initially or during follow-up, and for 55.6% required an immunosuppressive therapy, less frequently than AChRAB+ patients. MuSK, LRP4, and anti–striated muscle antibodies were not associated with cortactin antibodies. Also, in 201 AChRAb+ MG patients, 9.5% were cortactin antibody positive but they had the same clinical picture of classical AChRAb+ MG indicating that the presence of AChRAb prevails in prognosis terms. Cortactin antibodies have been proposed to be a biomarker of a good prognosis in patients with AChRAb- MuSKAb- MG and a useful biomarker in ocular MG without AChRAb or MuSKAb but its diagnostic/prognostic relevance remains to be proven [104, 283, 284].

# Chapter 9

# CYTOKINES, CRMP5, PRESYNAPTIC VOLTAGE-GATED CALCIUM CHANNEL ANTIBODIES OF THE P/Q AND N-TYPE, GLUTAMIC ACID DECARBOXYLASE 65 AND OTHER AUTOANTIBODIES

Autoantibodies to interferon –alpha2 (IFN-α2), IFN-ω and IL-12 have been reported in late-onset MG (LOMG) patients and in association with thymoma, especially in thymomatous myasthenic patients. These antibodies are mainly high-titer IgG with a neutralizing effect *in vitro* but their significance *in vivo* is still unknown. In a study on a great sample of patients [285], anti IFN-α2 have been found with ELISA in 32% of LOMG, in 57% of thymoma without MG and in 62% of thymoma with MG. These latter were positive in about 70% of cases before the immunosuppressive therapy and 56% after. Significant positivity in the same groups of patients was demonstrated also for anti IFN-ω and anti IL-12. Anti IFN and anti IL-12 antibodies seemed to be unrelated. Similar results were reported with RIA methodology [286]. No clear correlation was found with AChRAb titres and thymoma histology or other clinical features of patients but in some cases, a recurrence of thymoma was associated with a strong increase of antibody titres. Patients with viral infections, neoplasms or other autoimmune diseases and healthy subjects also can have cytokine autoantibodies [285].

Anti collapsin response-mediator protein-5 (CRMP5) antibodies are onconeural antibodies directed against a neural cytoplasmic phosphor-protein, associated with paraneoplastic syndromes, and have been found in 12-17% of thymoma patients with MG and in 7% in thymoma patients without MG; these antibodies are found also in 5% of patients with small cell lung carcinoma but at lower levels; CRMP5Ab are not correlated with the severity of MG but possibly with non-invasive thymoma [287].

Presynaptic voltage-gated calcium channel antibodies of the P/Q and N-type have been classically described in Lambert-Eaton syndrome but are also found in several other neurological syndromes, in association or not with a neoplasm (classically, small cell lung carcinoma), or very rarely in healthy controls [288]. These antibodies can rarely be found also in MG patients [289] or in MG-Lambert-Eaton overlap syndrome, with or without AChRAb seropositivity [290] but without thymoma [202]. Presynaptic voltage-gated calcium channel antibodies can be negative predictors for thymoma [291].

Glutamic acid decarboxylase 65 (GAD65, the enzyme that synthesizes GABA) antibodies are the main antibody specificity of the pancreatic islet cell in type 1 diabetes [292]. GAD65 Ab is detected generally at low levels in diabetes and in some healthy people (at ≥50 years of age) at higher levels in the stiff-man syndrome and other neurological syndromes. Also, GAD65 specificity was the most frequent neuronal antibody in a thymoma case series from the Mayo Clinic [202] with or mainly without MG.

Neuronal voltage-gated potassium channel (VGKC-complex) antibodies are described in association with peripheral nerve hyper-excitability (neuromyotonia), rarely accompanied by central nervous system hyperexcitability (Morvan syndrome). These antibodies also are reported in thymoma patients with MG, other neurological syndromes or without neurological symptomatology.

Other antibody specificities reported in thymoma patients with MG are: antineuronal nuclear antibody 1 (ANNA-1), anti ganglionic AChR, contactin-associated protein-like 2 (Caspr2), α-amino-3-hydroxy-5-methyl-4-isoxazolepropionic acid receptor (AMPAR), glycine receptor, and γ-aminobutyric acid-A receptor in a more recent case series from Mayo Clinic [291] and others. Neurological symptoms can also be absent and this neural autoantibody profile can aid in differential diagnosis of an anterior mediastinal mass.

In conclusion, thymoma is a strong initiator of the paraneoplastic immune response but onset of the clinical neurological syndrome is related to several other biological factors including titer and fine specificity of antibodies.

# Chapter 10

# BIOMARKERS RELATED TO B LYMPHOCYTES

## 10.1. ACTIVATED B CELLS AND PLASMA CELLS

A detailed analysis of the subtypes of B cell and plasmablasts/cells in MG patient groups and healthy controls showed reduced B cell frequency with decreasing CD27−IgD+ naive B cells after immunosuppressive drugs and with long-term disease [293]. Memory B cells were enhanced by immunosuppressive therapy. Another finding was the increased frequency of CD27 high plasmablasts/cells in all MG patients except for patients treated with steroids, who were similar to healthy controls. In ocular onset MG patients, enhanced frequencies of short-lived recently activated HLA DR high plasmablasts were evidenced. These could be a prognostic marker for secondary disease generalization and/or severity, but the design of the study was not suitable for detecting correlations of these markers with clinical severity. However, an increase of plasma cells in peripheral blood should suggest treatments directed against this target.

Another study suggested that monocytes/macrophages, dendritic cells (DCs) and B cells, expressing SDF-1 receptor CXCR4, are recruited from the periphery to the MG thymus, with a reduction of the expression of

CXCR4 on peripheral myeloid dendritic cells (also, the number of these cells decreased) [294]. Restoration of CXCR4 levels on peripheral myeloid dendritic cells and their numbers were accomplished by corticosteroid therapy.

In fact, B cells have been reported to fluctuate with therapy in MG. In refractory generalized MG, after rituximab therapy, reduced CD19+ B cells and CD19+CD27+ B cells parallels the clinical improvement of MG [295].

Other studies have been oriented to search markers related to B cells in MG.

CD21 is a major complement receptor expressed on B cells and has been investigated in the context of AChRAb production. Increased surface expression of CD21 on AChR specified B cells and decreased surface expression of CD21 on total B cells in the peripheral blood of patients with generalized MG have been reported by Yin et al. [296]. Moreover, the level of CD21+ AchR specified B cells correlated positively with serum anti-AchR IgG level.

Another potential B cell biomarker, CD72, a regulatory molecule that can induce hyperactivation or inactivation of B-cells [297] has been demonstrated to be altered in MG and multiple sclerosis [298]. In particular, the expression level of CD72 has a significantly negative correlation with AChRAb levels in MG.

These studies indicate that CD21 and CD72 may be involved in the pathogenesis of MG regarding to B-cell activation and AChRAb production.

As regards mechanisms maintaining the persistence of autoreactive cells, survivin has been indicated as a potential future marker. This protein inhibits apoptosis and consequently favors autoimmune cells survival, and has been already studied in several autoimmune conditions and in cancer. A small case study (15 MG patients and 10 controls subjects) focused on its role in MG pathogenesis and would support survivin as a marker of peripheral B lymphocytes and plasma cells for MG severity and response to treatment [299].

## 10.2. REGULATORY B CELLS [300]

Studies in mice have indicated that regulatory B cells have an important role for immune tolerance [300-307]. In MG patients, lower frequencies or defective regulatory B cells (Bregs) have been demonstrated as in some autoimmune disorders [308-313], and regarding their frequency and function seem to be inversely proportional to disease severity.

Regulatory B cells are difficult to study because of their rarity in the bloodstream and the lack of a known cell surface phenotype of identification: different research groups use different methods of analysis. However, production of interleukin-10 (IL-10) is a reliable measure of the suppressive function [301, 314, 315].

B10 cells are rare (about 0.6% of peripheral B cells in humans) and are capable of producing IL-10 for regulatory function on the T cell response [305, 316, 317]. In MG, a study demonstrated reduced frequencies of B10 cells in patients with anti-MuSK antibodies [312]. In addition, in another study, rituximab-treated MG patients with a rapid repopulation of B10 cells were associated with better outcomes [295].

A very recent study considered 64 patients affected by AChRAb MG from the Duke MG clinic (USA) and 21 healthy controls [300]. Disease severity was defined according to the MGFA Severity Class and MG-Manual Muscle Test. Detection of B10 cells was performed with a short-term stimulation with CD40L and lipopolysaccharide (LPS) or CpG along with restimulation with phorbal 12-myristate 13-acetate (PMA) and ionomycin [317]. After isolation of peripheral blood mononuclear cells, B cells producing IL-10 have been detected and IL-10 RNA and protein determined. When the entire group of patients were compared with controls there was not a difference. When patients were divided into groups with mild and moderate/severe disease basing on MGFA (or on a distinction between ocular and generalized MG), the frequency of B10 cells was inversely proportional to severity of MG. The independent induction of IL-10 by IL-21 and IL-35 was also demonstrated revealing that this effect was lower compared to toll-like receptor signaling by LPS and CpG. However, IL-21 and IL-35 could be used to enhance the generation of B10 cells holding in consideration that IL-21 can also enhance T cell response [318-322]. Next, the suppressive ability of B10 cells on CD4+T cell proliferation was demonstrated as depending on IL-10 but the suppressive function was lower in the moderate/severe group. In this study the most important result was that reduced frequencies of B10 cells are partially responsible for the loss of self-tolerance in patients with AChR antibody positive MG. The authors suggest that this can be a target of therapy [323, 324]. Another important consideration is that B10 cells could be used as a clinic biomarker of amelioration/worsening of MG, as suggested by the study of Sun et al. [295].

Another series of studies about Breg focused on T cell immuno-globulin mucin domain-1(TIM-1), a transmembrane glycoprotein on several cell types among which are T cells and dendritic cells, has been recently shown to be highly expressed also on B cells in an immune response *in vivo* [325]. Tim-1 identifies over 70% of all IL-10 B cells and thus is the most inclusive marker of this cell type [326]. Moreover, Tim-1 is essential for the regulatory activity of Breg and maintenance of self-tolerance [327, 328]. A recent study investigated the expression levels of Tim-1 on B cells from MG patients [329] and found that B cell Tim-1 expression levels are significantly lower in MG patients compared to healthy controls. Moreover, Tim-1 was negatively correlated with MG severity scores, plasma cells frequency, serum Th17-related cytokines and anti-AChRAb titers.

At present, B cell markers have not a role in clinical practice necessitating replication and standardization of unbiased laboratory data.

# Chapter 11

# BIOMARKERS RELATED TO T LYMPHOCYTES

## 11.1. T CELL RECEPTOR

The studies on T cell receptors (TCR) of MG patients have been conducted by FACS analysis and CDR3 spectratyping. With the first method Vβ5.1-, Vβ12-, and Vβ13- positive T cells [330-332] were reported to be significantly increased in MG patients. The more sensitive CDR3 spectratyping found contrasting results [333, 334]. A more recent CDR3 spectratyping study on peripheral blood lymphocytes and thymus tissues from MG patients has not revealed preferential usage of T cell receptors but longitudinally persistent oligoclonal expansions of Vβs, which differed from patient to patient. These latter seemed to correlate with clinical severity and high AChRAb titer [335].

More recently, peculiar CD8+TCRVβ-subset expansions were found in the majority of LOMG or thymoma-associated MG [336]. Some of these expanded cells showed a naïve CD62L+hi/CD45RA+recent thymic emigrant (RTE)-like phenotype and could be associated with IgG against cytomegalovirus, IL-12 and/or IFN-α2.

## 11.2. T HELPER CELLS

CD4+ T helper (Th) cells with reactivity against AChRs activate B cells to produce specific antibodies by producing pro-inflammatory cytokines and delivering costimulatory signals [337]. CD4+ Th cells are classified into 4 major subsets, Th1, Th2, Th17 and regulatory T cells (Treg) on the basis of cytokine production and expressed transcription factors. The Th1 cells are pro-inflammatory cells producing interferon-γ (IFN-γ) and interleukin-2 (IL-2), which can transfer cell-mediated immunity. The Th2 cells secrete anti-inflammatory cytokines like IL-4 and IL-10, and can also stimulate B cells to produce antibodies. The Treg cells are considered to be anti-inflammatory. The Th17 with their main cytokine IL-17 promote inflammation and autoimmunity. These types of T cells differentiate from naïve CD4+T cells with IL-1β. This latter induces retinoic acid–related orphan nuclear hormone receptor γT (RORγT) expression and Th17 differentiation, which is amplified by IL-6 and IL-23 [338, 339]; IL-23 is essential for the full development and survival of Th17 cells. Interleukins-1 and 23, derived from activated dendritic cells, can induce IL-17 from γδT cells [340]. Also, IL-23 plus IL-1β and prostaglandin E2 can induce the Th17 immune response in CD161+CD4ι memory T cells in inflammatory bowel disease [341].

Activated Th17 cells produce proinflammatory cytokines as IL-17A, IL-17F, IL-21, IL-22 and TNF-a [342-345].

Th17 are the major pathogenic T-cell subsets in experimental autoimmune encephalitis [346] and in several human auto-immune diseases such as rheumatoid arthritis, systemic lupus erythematosus, inflammatory bowel disease and multiple sclerosis. Moreover, IL-17 is a major mediator of tissue inflammation [347] and upregulates several chemokines and matrix metalloproteases, with the recruitment of neutrophils [347].

In MG patients, Th17 cells influence the Th1/Th2 cytokine balance in PBMNCs and stimulate the production of autoantibodies [348].

In the process of the breaking of self-tolerance, Th-17 cells would have a paramount importance and indeed, transgenic IL-17 null mice are resistant to the induction of EAMG: B-cell autoimmunity in EAMG is dependent on CD4+ T cells expressing IL-17 [349].

The importance of Th17 cells has also been demonstrated in another EAMG model, the null mutant mice for the genes of both the IL-12/IL-23 p40 subunit and IFN-gamma, a model in which Th1 function is impaired, and thus Th1 pathogenic role is excluded, in the presence of a normal function of pathogenic Th17 cells [350].

In EAMG, IL-17 and Th17 cells have been shown to coordinate autoreactive T and B cells permitting antibody production: in the late phase of the disease Th17 ratio increases and IL-17 is upregulated [351].

In a clinical study, the frequency of Th17 cells was similar in subjects with and without MG [352]. However, there was a significant correlation between Th17 cells and the changing rate of AChRAb titer (%) and a negative correlation between the relative RORγT mRNA levels and the Th1 / Th2 ratio in CD4+ cells.

Another clinical study demonstrated that the serum concentration of IL-17 is significantly increased in generalized MG compared with ocular MG and controls, correlates with the severity of the disease and with AChRAb titers [353], but the authors believed that these findings needed to be confirmed in a larger sample including more patients with thymic alteration.

Other researchers investigated the role of Th17 cells in the pathogenesis of MG [354]. Eighty six patients with MG and 32 healthy controls were studied. Patients were divided in 35 MG with thymoma, 30 MG with thymic hyperplasia and 21 MG with normal thymus. The Th17 cell population and serum levels of related interleukins (IL-17, IL-1β and IL-23) were significantly increased. The Treg cell population was decreased in peripheral blood mononuclear cells (PBMCs) of MG patients with thymoma. The increase of Th17 cells correlated with the severity of MG as determined with the QMG score in patients with thymoma but not in patients without. A correlation was found between the frequency of peripheral Th17 cells and AChRAbs of patients with MG. However, the AChRAb concentration has no relationship with the subtype of MG. In 10 MG patients with thymoma, after thymectomy, the Th17 cells tended to decrease but without significance. In MG Th17/Treg imbalance and Th17 cytokines have a crucial role especially in the immune process of MG in the context of thymoma.

Other recent studies support the role of Th17 in MG.

The tumor necrosis factor (TNF)-α-induced protein 8-like-2 (TNFAIP8L2 or TIPE2) is a negative regulator of the TLR4-mediated autoimmune Th17 cell response. A recent study found that the expression of TIPE2 was reduced in the peripheral blood mononuclear cells (PBMCs) of MG patients compared with normal controls, more in generalized forms than in ocular ones [355]. Furthermore, a significant negative correlation with the serum levels of IL-6, IL-17 and IL-21 was evidenced. Consequently, TIPE2 may participate in the development of MG and could be studied as a putative marker of the disease.

Another study considered the P2X7receptor (P2X7R), an ATP-gated cation channel, expressed in various immune cells [356]. This receptor has a pro inflammatory function stimulating cytokines as IL-1β and IL-6 [357-359] promoting Th17 immune response.

Levels of P2X7R in PBMCs from 32 MG patients (12 generalized and 20 ocular MG) and 22 healthy controls were determined. The expression of P2X7R was found to be increased in MG patients with positive correlation, clinical severity and serum levels of Th17 related cytokines. The P2X7R is involved in the pathogenesis of MG and could be studied as a putative marker of the severity of disease.

B7 is a family of surface proteins involved in inflammation and immunity; for example, B7-1 on the surface of APC binds to CD28 or to CTLA-4 on T cells providing a costimulatory or coinhibitory signal in the context of T cell receptor activation. B7-H3 (B7 homologue 3, CD276), a member of the B7 family, can be expressed (it is not constitutively expressed, but is induced by inflammatory cytokines) on T cells, natural killer cells, dendritic cells, and monocytes. Receptor(s) of B7-H3 are so far unknown (some data would indicate TLT2) [360], however they are expressed on activated T cells. According to the literature, B7-H3 has a prevalent coinhibitory function on T cells with a partial costimulatory action reported by some studies [361]. Soluble B7-H3 and membranous

B7-H3 provide inhibitory signals to T cells [362, 363]. The expression of B7-H3 on peripheral blood mononuclear cells and its soluble form have been evaluated in MG patients [364]. Significant differences have been found between MG patients and healthy controls, in generalized versus ocular MG, in abnormal versus normal thymus and according to QMGS revealing a fine down-regulation of soluble B7-H3 correlated to the severity of MG. However, the B7H3 gene had not been demonstrated to be associated with MG [365].

A study signaled an overexpression of OX40 on CD4+ T-cells from MG patients associated with high levels of AChRAb, thymic hyperplasia and onset at an early age [366]; elevation of OX40+ CD4+ T cells, followed by the elevation of CD4/CD8 ratio, preceded the onset of MG in the context of chronic graft-versus-host disease after allogeneic bone marrow transplant [367].

A series of studies considered fluctuations of T cell subtypes in response to treatment.

After double filtration plasmapheresis, the percentages of T-cells, Th cells, and the Th/Ts ratio decrease, whereas the percentage of NK cells increase [368]. Quantitative MG scores decreased in correlation with percentage of T cells.

In MG patients, a significant correlation between T helper type 1/type 2 ratio and the P-glycoprotein function (a drug efflux pump) on CD3+T cells was demonstrated [369]. With prednisolone treatment, the percentage of change in the T helper type 1/type 2 ratio and the reduction rate of quantitative myasthenia gravis scores after 12 months of treatment were negatively related. With prednisolone and calcineurin inhibitor, MG patients ameliorated regardless of the T helper type 1 predominance. Supplemental calcineurin inhibitors should be useful in MG patients treated with prednisolone when their T helper balance shifts toward type 1.

A study suggested a significant correlation between P-glycoprotein activity, a drug efflux pump that actively transports glucocorticoid out of cells, on CD3(+) or CD4(+) cells and steroid doses to achieve a clinical response [370]. The measurement of P-glycoprotein function on CD4(+) T cells could be useful in the evaluation of a clinical response to glucocorticoids.

## 11.3. FOLLICULAR CD4+ TH CELLS

A recent review summarized data about follicular CD4+ Th cells (TFH cells) in MG [371].

TFH cells, present both in lymphoid tissues and circulation, stimulate antibodies generation by B cells and are implicated in the pathogenesis of MG: frequencies of CD4+CXCR5+ [372, 373], CD4+CD45RO+ CXCR5+, CD4+CXCR5+PD-1hi, and CD4+CXCR5+ICOShi T cells in the peripheral blood from MG patients was higher compared to healthy subjects. A significant positive association between the CD4+CXCR5+ T cells and disease severity was also reported [372, 373]. Moreover, the frequencies of CD4+CXCR5+ICOShi or CD4+CXCR5+PD-1hi T cells were positively associated with serum AChRAb titers antibody. The percentage of CD4+CXCR5+ T cells was decreased after treatment. Levels of serum CXC chemokine ligand 13 (CXCL13, a chemokine that can provoke abnormal migration of immune cells and, consequently, augment pathogenic immune reactions in lymphoid tissues) was increased in MG [374, 375] and positively correlated with disease severity and circulating CD4+CXCR5+ICOShi T cells [373]. IL-21, important for germinal centers formation, TFH cell differentiation, and antibody production, is increased and positively related to the percentage of CD4+CXCR5+ICOShi T cells in MG [373].

In experimental autoimmune myasthenia gravis CD4+CXCR5+PD-1+ TFH cells and TFH-associated molecules, Bcl-6 and IL-21, in mononuclear cells of spleen, have been shown to be upregulated and AChRAb titers were associated with the frequency of TFH cells in spleen [376]. Moreover, inhibiting Bcl-6 in EAMG ameliorated clinical severity, reduced TFH cells, Bcl-6, IL-21, and AChRAb titers [376]. CD4+CXCR5+ T cells could be potential markers to assess the MG activity and the effects of treatments.

## 11.4. CD4+CD25+ REGULATORY T CELLS

In peripheral blood, CD4+CD25+ regulatory T cells (Treg) have a suppressive function on several types of hematopoietic cells and mainly on T effector cells. Dysregulation of Treg have been implicated in the pathogenesis of inflammatory and autoimmune diseases such as systemic lupus erythematosus, rheumatoid arthritis, multiple sclerosis and also myasthenia gravis [377]. The transcription factor, forkhead box protein P3 (FOXP3) is expressed in CD4+ CD25+ Treg cells and is the principal regulator for the development and function of these cells that play a critical role in maintaining self-tolerance as well as in regulating immune responses [378, 379]. Several mechanisms of the suppressive function of Treg have been reported among which direct cell-to-cell contact, cytokines production, adsorption of IL-2 by CD25++ [380].

The IL-2 signaling pathway stimulates the FOXP3 expression in Treg by activation of signal transducer and activator of transcription-5 (STAT5) [381, 382]. Other Treg markers are glucocorticoid-induced tumor necrosis factor receptor-related protein (GITR) and cytotoxic T-lymphocyte antigen-4 (CTLA-4) [383]. The inhibitory pathways of programmed death-1 (PD-1) and its ligands (PD-L1 and PD-L2) as well as the GITR interaction have also been implicated in the mechanism of action of Treg [384, 385]. Both PD-1 and PD-L1 are present on Treg. However, a study denied association of the PD-1 gene to MG, although admitted an increased expression of PD-1 and its ligand PD-L1 on CD4+ T cells [386]. The active form of Vitamin D, 1.25-dihydroxyvitamin D3 (VitD3) has been shown to increase the percentage of CD4+FOXP3+ Treg [387, 388]. Plasma VitD levels have been shown to be lower in MG patients [389].

According to several studies the suppressive capacity of Treg is impaired in AChRAb+ MG patients [377, 390, 391]. Both primary intrinsic defects of Treg [377] and deficits of responding and suppressing cell types have been reported [392].

CD4+ CD25+ Treg cells and TGF-beta1 have been shown to be significantly decreased in MG patients and conversely correlated with levels of anti-AChR Abs in MG patients without thymoma [393].

Another study demonstrated in MG patients decreased CD4(+) CD25(high)Foxp3(+) regulatory T cells and increased CD19(+)BAFF-R(+) B cells [394]. Serum ICAM-1 was increased as reported also by others [395]. These latter authors demonstrated increased serum levels of CD25 in MG patients.

A study has been conducted to clarify the dysfunction mechanisms in Treg in 78 generalized, AChRAb+ MG patients, divided in 52 EOMG patients and 26 LOMG [67]. Sixty percent of patients were thymectomized and none had thymoma. Fifty nine percent of the patients were receiving immunosuppressive (IS) therapy with steroid or steroid plus azathioprine. Sixty-one healthy controls were also studied. A high expression of CD25 was used to identify Tregs, as already done in other studies [396, 397].

EOMG and LOMG patients were not significantly different; PD-L1+ CD4+CD25++ cells were reduced, whereas PD-1+ Tresp were increased mainly in LOMG; decreased FOXP3 and STAT5 phosphorylation were associated with the dysfunction of Treg of AChRAb+ MG patients on co-culture with Tresp.

The dysfunctional state of Treg in patients with both early- and late-onset AChRAb+ MG was thus confirmed and can be related to an unstable interaction between Treg, Tresp, and APC considering various co-stimulatory/co-inhibitory molecules. The decrease of FOXP3 expression in Treg may be caused by decreased IL-2 mediated STAT5 phosphorylation. Finally, VitD3 has a therapeutic potential modulating Treg-related suppression in AChRAb+ MG.

In children, Tregs have been shown to be increased in number but functionally unable with low Foxp3 and CTLA4 [398].

Another study [399] investigated peripheral Tregs in childhood ocular MG. Thirteen children with AChRAb+ ocular MG and 18 age-matched controls were analyzed. The percentage of Tregs in CD4+T cells did not depend on age, values were comparable between children and adults. The Treg population was significantly lower in the active stage than during remission (following immunosuppressive therapy) and in comparison with the control group (percentages of Tregs in peripheral blood CD4+ T cells in active stage, remission stage, and control groups was 3.3 ± 1.3%, 4.8 ±

1.7%, and 5.0 ± 0.6%, respectively). Furthermore, the Treg percentage was significantly lower during relapse of myasthenia symptoms. No associations between the percentage of Tregs and immune suppressant dosages were found. Moreover, peripheral blood Th17 cells did not appear to be related with disease activity. The authors concluded that the Treg population was significantly reduced contributing to the pathophysiology of ocular type childhood MG and that could be a marker of the immunological state in these patients.

Serum levels of T regulators have been evaluated in nonthymomatous MG patients after thymectomy. Higher percentages of Treg with significantly better MG outcomes were obtained if immunosuppressive therapy was associated with thymectomy [400].

Several studies on thymic tissue seem to confirm the importance of Treg in MG pathogenesis. A study investigated the thymic stromal lymphopoietin (TSLP) in MG with thymoma [401]. TSLP is an IL-7 like cytokine acting on several types of cells such as dendritic cells, CD4+, CD8+ and natural killer (NK) T cells, B cells and epithelial cells [402]. Really, TSLP is an important regulator that can stimulate the generation of Treg and decrease the differentiation of Th17 cells [403]. Interestingly, TSLP is aberrantly expressed in various tumor tissues (for example, breast and pancreatic cancer) [404, 405]. In thymomas, TSLP is significantly decreased because it is post-transcriptionally downregulated by a microRNA, miR-19b-5p, which is overexpressed. Results of this study indicate that TSLP and miR-19b-5p could have a crucial role in the pathogenesis of MG associated with thymoma. Moreover, according to authors, miR-19b-5p could be considered a potential biomarker for thymoma and MG.

In conclusion, T cell markers could be a biomarker for disease severity/response to treatment but the studies are only at the start [406].

# Chapter 12

# INFLAMMATORY PROTEINS

Myasthenia gravis is a T cell dependent and B cell mediated disease in which a complex interplay of cytokines modulates the immune response and is implicated in pathogenesis. Several studies investigated the role of specific cytokines as biomarkers of MG. In particular, IL-17 has been the object of interest because of its importance for survival and proliferation of human B-cells and their differentiation into immunoglobulin-secreting cells [407]. In experimental autoimmune MG in rats, Th17 cells (a CD4+ helper T-cell subset related to inflammation and autoimmunity) and IL-17 were shown to be upregulated [408]. This cytokine has been demonstrated to be significantly increased in generalized MG patients compared with ocular MG patients and healthy controls; moreover, IL-17 levels correlated with disease severity and AChRAb titers [407]. Another study confirmed the upregulation of IL-17 and added that IL-22 was decreased in MG patients in comparison with healthy controls [409]; MG PBMC treated with IL-22 showed a depressed production of IL-17: all these data led to consider IL-22 as a protective factor. Moreover, as the decreased IL-22 mRNA levels were found in patients with generalized MG with respect to those with ocular MG, IL-22 was suggested by the authors as a potential marker of the severity of MG.

Interleukin-2 is overproduced by T cells of MG patients and is partly responsible for their proliferation and their relative resistance to inhibitory effects of mesenchymal cells, a potential therapy for autoimmune diseases [410].

The great majority of studies on cytokines regard adult patients. A study on pediatric patients with ocular MG found that IL-17A levels were not different from those of healthy controls [411]. The same has been observed for IL-6 and interferon γ. However, levels of these cytokines became undetectable after immunosuppressive therapy. Conversely, levels of IL-2, IL-4, IL-10, and TNF-α were below detectable levels before and after immunosuppression.

In adults, elevation of IL-21 (produced by TFH and Th17, determining B cell differentiation into plasma cells) and IL-27 (produced by activated antigen presenting cells and previously described as enhanced in inflamed tissues), have also been reported in MG by other authors without correlation with AChRAb titers [412]. A very recent study found that IL-21 amplifies T cell responder proliferation *in vitro*, in the presence of Treg, through IL-2 inhibition: consequently, a therapy with a low dose of IL-2 could be utilized in MG but other studies are necessary to provide evidence for it [413].

Haptoglobin is an acute phase protein and its production is stimulated mainly by IL-6 in the liver. In acute inflammation, haptoglobin would have an inhibitory function, in fact it acts as an immunosuppressor of lymphocyte function and modulates the Th1 and Th2 balance and moreover inhibits the CD22 binding to activated TNFα on the surface of B cells. In a study, haptoglobin has been suggested to be a marker of clinical activity in MG without another associated illness (it would reflect the activity of any inflammatory disease) [414].

A study [76] determined plasma levels of cytokines related to various Th subtypes in AChRAb+ and MuSKAb+ MG patients and healthy controls (HC). Plasma levels of cytokines related to Th1, Th2 and Th17 were not significantly different between MG subtypes and healthy controls. The study of the stimulation of PBMC from MuSKAb+ MG but not AChRAb+, MG patients showed significantly increased secretion of the

Th1, Th17 and T follicular helper cell related cytokines, IFN-γ, IL-17A and IL-21. Immunosuppression reduced levels of IL12p40 in the plasma of AChRAb+ MG and MuSKAb+ MG patients, decreased spontaneous secretion of IFN-γ and stimulated secretion of IL-6 and IL-10 by PBMC from AChRAb+ MG, but not MuSKAb+ MG patients. The authors concluded that Th1 and Th17 immune reactions are very important in MuSKAb+ MG. Immunosuppression decreases the Th1 response in AChRAb+ MG and MuSKAb+ MG, but in other respects modulates immune responses in these form of MG differentially.

The frequency of CD4+ T cells expressing Th1 type chemokine receptor CXCR3 (receptor for some proteins among which IFN-gamma-inducible protein of 10 kDa, IP-10) was found to be reduced in untreated AChRAb+ MG patients, mainly in thymoma MG, with a significant inverse correlation between the percentage of these cells and the disease severity [415]. The frequency of CXCR3-positive CD4+ thymocytes from the thymomas and hyperplastic thymuses was found to be increased. Long after therapy (thymectomy and corticosteroids), the frequency of CXCR3-positive CD4+ T cells normalized. The frequency of CXCR3-positive CD4+ T cells was correlated with the elapsed months after thymectomy. Consequently, positivity for CXCR3 was proposed as a potential marker of the long-term disease activity. A preceding study evidenced an increase in IP-10 and its receptor, CXCR3, in the thymus and muscle of MG patients whereas CXCR3 was overexpressed on the surface of their peripheral CD4+ T cells [416]. This overexpression of CXCR3 was present also in some patients of Suzuki et al. [415].

The B-cell activating factor (BAFF) is a 285-amino acid long glycoprotein with a cytokine function that belongs to the TNF ligand family and is secreted by macrophages, dendritic cells, and neutrophils [411]. This cytokine has a strong immunostimulant activity on B cells and its overexpression can lead to immune diseases such as SLE. Serum BAFF levels have been demonstrated to be significantly higher in adult AChRAb+ MG patients [417] and in childhood onset ocular MG [411] in comparison with controls, decreasing after immunosuppressive therapy and significantly correlating with AchRAb levels [418]. A role of BAFF as biomarker of therapeutic effect has been proposed. Significantly elevated levels of BAFF decreasing with immunosuppressive therapy are confirmed by other studies, without correlation to MG severity and thymectomy [419]. A recent study confirmed increased BAFF levels also in MuSKAb+MG patients [312].

Relatively old studies [420] demonstrated that cytolytic T lymphocyte-associated antigen-4 (CTLA-4) is differently expressed in MG with low surface and intracellular levels of CTLA-4 protein and high levels of soluble form in serum. The latter is positively correlated with the serum concentration of AChRAb. A more recent study about genetic polymorphism of CTLA4 is discussed in genetic biomarker chapter.

Proteomic studies considered serum protein profiles in MG searching for disease biomarkers. A study found peculiarities in proteomic spectra distinguishing MG patients from healthy controls [421].

Very few studies investigated a broad inflammatory circulating protein profile in MG.

One study [74] showed significantly increased serum levels of a proliferation-inducing ligand (APRIL), and cytokines IL-19, IL-20, IL-28A and IL-35 in AChRAb+ MG as compared with healthy controls. Moreover, IL-20, IL-28A and IL-35 were significantly decreased after immuno-suppression. APRIL is secreted by macrophages, T cells and dendritic cells and favors maturation and survival of B cells [422]. Interleukin-19 can be considered an anti-inflammatory cytokine, IL-20 pro-inflammatory, IL-28A and IL-35 can be considered anti-inflammatory [74].

Another study that considers the inflammatory circulating protein profile in MG is that of Molin et al. [75], in which 92 different proteins, inflammation related, were investigated within the clinical subgroups EOMG vs. LOMG, immunosuppressive medication, gender and thymectomy. The analysis kit used was the OlinkProseek Multiplex Inflammation I 96×96 kit (Olink Bioscience, Uppsala, Sweden). Forty-four MG patients (22 EOMG and 22 LOMG; 39 AChRAb+ and 5 AChRAb-MuSKAb-) and 43 healthy controls were evaluated. The median score of MGFA classification was mild (class II).

The more significantly increased proteins were matrix metalloproteinase 10 (MMP-10), transforming growth factor alpha (TGF-α) and extracellular newly identified receptor for advanced glycation end-products binding protein (EN-RAGE) (p < 0.01). Other proteins reaching the statistical significance were nerve growth factor beta (β –NGF), cytokines IL-6, IL-8, IL-10 and IL-17A and C as well as chemokines C-C motif ligand 19 (CCL19) and C-X-C motif ligand 1 (CXCL1). Early onset MG patients have significantly higher levels of MMP-10 and CXCL1 in comparison with LOMG patients whereas the latter have higher levels of BDNF.

Matrix metalloproteinase 10 degrades the extracellular matrix both in physiological and disease processes. In experimental autoimmune encephalomyelitis (a multiple sclerosis model similar to experimental autoimmune MG) MMP-10, secreted by microglia and macrophages, has been shown to be elevated, correlating with disease severity [423].

TGF-α is a growth factor produced in macrophages, neurons, astrocytes and keratinocytes that stimulates proliferation and different-iation of epithelial and neural cells but has not known roles in the peripheral nervous system [75].

EN-RAGE, among other its properties, inhibits metalloproteinases and has pro-inflammatory effects on lymphocytes, neutrophils and endothelial cells [424]: is induced by TNF-α and IL-6, that are involved in the pathogenesis of MG [76, 425, 426], and induce several cytokines.

β-NGF, besides its known role in central and peripheral nervous system, is released by T and B cells and involved in inflammation [427].

IL-6 and IL-10 are produced by B and T cells and act on these; in particular, IL-6 stimulates Th17 cells that produce IL-17 (see the above mentioned effects of this interleukin and its importance for MG).

Interleukin-8 and CXCL1 are pro-inflammatory cytokine upregulated in MG [428].

CCL19 is a cytokine secreted by medullary thymic epithelial cells and upregulated in the thymic hyperplasia [429].

BDNF is a neurotrophic factor that is decreased by pro-inflammatory cytokines [430] and possibly has higher concentrations in LOMG because this latter is less driven by inflammation in comparison with EOMG.

However, the authors underline that proteins levels are expressed in arbitrary units and this problem limits comparisons between proteins and also for the same protein determined with different assays.

In a study on immunoadsorption in MG, IL-18 and TGF-β were shown to be decreased and IL-10 was shown to be increased after treatment [431].

The complement system is activated in AChRAb+ MG by three complement activation pathways, classical, alternative, and lectin, which converge at the assembly of C3. C3 levels have been determined in AChRAb+ MG patients and have been found to be inversely correlated with severity of AChRab+ generalized MG as measured by the quantitative MG score [432] and so useful for assessing therapeutic effects. In some publications, circulating immune complexes had been related to the severity of experimental MG [433] and similar data had been anecdotally reported in human disease [434].

Mannose binding lectin (MBL) is a lectin (a carbohydrate-binding protein) belonging to the collectin (collagen-containing C-type lectins) family, an acute phase protein produced in the liver in response to inflammation. This protein can activate the complement system with the lectin pathway starting with the binding of mannose, glucose or other sugars placed on surface of microorganisms, without the reaction antigen-antibody. Although the complement system has an undeniable importance in MG pathogenesis, Li et al. [435] found that MBL serum levels are no significant different between AChRAb positive generalized MG patients and healthy controls. Moreover, they showed that MBL gene deficient mice could develop the disease condition after immunization with AChR. In fact, they denied a role for a MBL pathway in MG. Nevertheless, other authors [436] demonstrated significantly increased plasma levels of MBL in 30 MG patients compared to healthy controls. Furthermore, MBL levels were significantly correlated with MG severity. These findings suggest that MBL contributes to the pathogenesis and also severity of MG. The use of

MBL as a biomarker of disease awaits confirmations in a greater number of patients.

Human resistin is a 12.5 kDa peptide hormone, rich of cysteine (10-11 cysteine residues), consisting of 108 amino acids, implicated in insulin resistance, immune response and inflammation [437]. Initially described in mice adipocytes [438], resistin was later reported in humans in monocytes/macrophages, neutrophils and lymphocytes [439].

Resistin induces the expression of various cytokines, including tumor necrosis factor-α (TNF-α), interleukin-1 (IL-1), interleukin-6 (IL-6) and interleukin-12 (IL-12) [440] and these effects are mediated through the NF-κB (nuclear factor kappa-light-chain-enhancer of activated B cells-mediated) signalling pathway. Resistin itself is induced by IL-1β, IL-6 and TNF-α [441, 442] suggesting that resistin can induce itself with a positive feedback mechanism [440, 443]. Moreover, resistin exerts chemotactic activity on CD4+ lymphocytes [444] through intercellular adhesion molecule-1 (ICAM1), vascular cell-adhesion molecule-1 (VCAM1) and chemokine (C-C motif) ligand 2 (CCL2) [445].

Resistin is reported to be related to inflammation in rheumatoid arthritis, systemic lupus erythematosus, and idiopathic inflammatory myopathies [446-448].

Regarding rheumatoid arthritis, an animal study has demonstrated that resistin injected in articulations can induce leukocytes infiltration, as it happens in human RA [449]. Moreover, serum resistin levels correlate with disease activity and levels of acute phase proteins suggesting that resistin may be a mediator of inflammation in RA [450-452]. Resistin levels have been shown to be correlated to lymphocytic inflammation in the salivary glands of patients with primary Sjogren's syndrome [453].

A study described serum resistin levels in AchRAb MG [454]. One hundred and two AchRAb+ MG patients of the Chinese Han ethnic group, that were not practicing immunosuppressive treatment, were studied, comprising ocular and generalized forms. MG-activities of a daily living (MG-ADL) score was used to try and correlate serologic and clinical data. Acetylcholine receptor antibodies status was determined with cell based assay [455] and scored semiquantitatively [456]. Resistin levels were measured using the ELISA procedure. Levels of resistin were significantly higher in AChRAb+ MG patients compared to controls; patients with generalized disease had significantly higher resistin levels than patients with an ocular form and these were not significantly different from controls; patients with MG with thymoma had the highest resistin levels. On these last data, it is necessary to add that in literature, resistin has been shown to be involved in the pathology of malignant tumors [457, 458]. A significant relationship between serum resistin levels and MG-ADL scores was also reported in AchRAb generalized MG patients and in MG patients with thymoma; no significant correlation was found between resistin levels and anti-AChR antibody scores. Six generalized MG patients were also studied before and after immunosuppressive treatment and serum resistin levels were lower after treatment than in patients not receiving therapy. This final data is not surprising because decreased resistin levels have been shown to be a marker of successful therapy in patients with inflammatory bowel disease [459, 460].

The results with the data on other immune diseases in literature have been considered by the authors to be suggestive of an important role of resistin in the MG pathogenesis and its possible use as biomarker for MG severity and the association with thymoma [454]. Recently, the same authors confirmed their conclusions [461]. On the contrary, other authors from Brazil [462] found decreased resistin levels in their MG patients and attributed this difference to glucocorticoid treatment: in fact, Chinese patients were glucocorticoid naïve whereas all Brazilian patients used this therapy.

Heat shock proteins (HSPs) are proteins present in nearly all nucleated cells, induced by inflammation and other stressors, classified according to their molecular weight. These proteins act as molecular chaperones and inhibit caspase-dependent (e.g., HSP27, HSP70) and -independent apoptosis pathways (e.g., HSP70). A study considered HSP27 and 70 in serum and thymic tissue of patients with thymoma and thymic hyperplasia, serum levels were also evaluated in controls [463]. Serum HSP27 and HSP70 concentrations correlated significantly with the WHO tumor classification and were more elevated in thymoma patients, mainly in thymic carcinoma patients in comparison with healthy volunteers and patients with thymic hyperplasia. Serum HSP27 concentration was significantly higher in patients with Masaoka stage III–IV compared to stage I–II tumors and significantly decreased after complete tumor resection. No significant HSP differences were found between thymoma patients with or without MG, but MG patients showed higher serum concentrations of HSP27 compared to volunteers.

Antibodies to HSP-65 have been described in 80% of 40 MG patients [464]. According to authors, these antibodies can be derived from an increased expression of HSP-65 and could have a role in the pathogenesis of MG.

The Fas receptor (or CD95) is a surface receptor that belongs to the tumor necrosis factor family. This receptor, activated by a ligand, determines cell apoptosis. Peripheral T cells of CD4+, CD4+CD8- and CD4-CD8- subsets of MG patients were found to have upregulation of Fas antigen [465] with significant differences between MG patients with normal thymus and MG patients with thymoma, MG patients with and without hyperthyroidism, and during corticosteroid treatment, but no difference between patients with ocular and generalized MG. Authors concluded that Fas antigen might be involved in the pathogenesis of MG, in the mechanisms of corticosteroid treatment, and in hyperthyroidism comorbidity. Ocular MG may represent a systemic disease.

The upregulation of several phosphodiesterase subtypes in muscle and lymph node cells of rats with experimental autoimmune MG was established determining mRNA expression [466]. The overexpression of phosphodiesterases seemed to be correlated to MG severity. Similar overexpression was confirmed by the determination of mRNAs in thymus and peripheral blood lymphocytes of AChRAb + generalized MG patients. The same conclusions were reached for human multiple sclerosis and its animal model. The authors proposed phosphodiesterase mRNA determination as biomarker of MG severity.

Toll-like receptors (TLR) are molecules with receptor function (at present, 13 types of TLRs are known in mammals) involved in innate immunity as pattern recognition receptors and adaptive immunity because promote the maturation and activation of antigen-presenting cells [467]. Levels of mRNA of TLRs in PBMC of MG patients and controls were determined with quantitative reverse transcription polymerase chain reaction and compared with quantitative myasthenia gravis score. Aberrant expression of TLRs and overexpression of TLR 9, correlated with MG clinical score, were discovered suggesting a possible involvement of TLRs in MG pathogenesis.

A study considered the metabolomic profile of MG patients before and during chronic corticosteroid treatment in search of serum biomarkers that can successively be validated for clinical efficacy or adverse effects [468]. The upregulation of membrane associated glycerophospholipids was observed, which may be associated with certain adverse effects: deviations in phosphatidylcholine levels in serum have been associated with early markers of risk of cardiovascular disease [469]. Arachidonic acid and its derived pro-inflammatory eicosanoids were reduced: these could be markers of response to treatment. Other perturbations were observed among which an increase of Vitamin E succinate and decrease of 3-Deoxyvitamin D3. The authors defined their work as a "proof of principle" needing confirmations and validation for clinical purposes.

In conclusion, all the proteins shown to be significantly increased/decreased in comparison between patients and controls or between specific subgroups of patients might be potential biomarkers of inflammation and response to treatment in MG but methodological problems and the need for replication of data at present preclude the practical use.

# Chapter 13

# SERUM ALBUMIN, URIC ACID AND OTHER ANTIOXIDANTS

Oxidative stress contributes to the pathogenesis of inflammatory and autoimmune diseases [470] and in particular of neuromuscular diseases [471]. Reactive oxygen species might damage the AChR [472]. Highly conserved cysteine residues would be the major targets of reactive oxygen species in nicotinic AChR [473, 474].

Some antioxidants naturally occurring in serum have been investigated to determine their relevance in the pathogenesis of MG.

Human serum albumin (HSA) has several functions among which its strong antioxidant capacity. This protein contains 585 amino acids and has a molecular weight of 66 kDa. It is a monomeric multi-domain macro-molecule, the most abundant protein in plasma (at normal concentrations between 35 and 50 g/l) and the main determinant of plasma oncotic pressure. Moreover, it is a carrier for many compounds (metal ions, fatty acids, cholesterol, bile pigments, drugs, hormones), binds to potential toxins making them harmless, and is a strong anti-oxidant.

Regarding this latter property, HSA acts through its multiple-binding sites trapping free radicals. Preceding studies have demonstrated that more than 70% of the free radical-trapping capacity of serum was due to human serum albumin (HSA) as assayed using the free radical-induced hemolysis test [475, 476]. In physiological and pathological conditions, the structure of HSA can be altered obtaining the oxidized form with a loss of beneficial antioxidant properties. Oxidized albumin (the ratio of the oxidized form to the normal form of albumin) is a biomarker of oxidative stress [477].

At present, HSA has been implicated in the pathogenesis of diseases with inflammatory and free radical damage. Indeed, HSA is a valuable biomarker of rheumatoid arthritis, renal disease and graft-versus-host disease [478].

Actually, levels of HSA decrease with inflammation. Renal disease has been especially studied in this regard. The reduction of HSA in early stages of kidney diseases is associated with inflammation and increased risk of cardiovascular diseases. In one study, serum CRP level >0.6 mg/dl was associated with a decrease in HSA level by 70 mg/dl (95%CI, 0.03 to 0.12) [479]. In end stage renal disease, characterized by an inflammatory state, high levels of ESR, hepcidine, and ferritin and low levels of HSA, LDL and HDL cholesterol may be revealed [480-486]. Also in these situations, serum CRP may be related to HSA levels which are affected by an inflammatory response [487, 488].

In another study of 757 hemodialysis patients followed up for 30 months, the relationship between hs-CRP, IL-6, IL-8, and HSA with mortality and morbidity was determined by multivariate analysis [489]. The laboratory values were strong predictors of mortality. In particular, high serum CRP and pro-inflammatory cytokines and low HSA levels were related to the highest risk for cardiovascular and all-cause mortality [489].

Researchers of the University of Wenzhou (China) have studied HSA in MG [490]. One hundred sixty-six MG patients from the First Affiliated Hospital of Wenzhou Medical University were enrolled. The disease severity assessment was calculated with the MGFA clinical classification. Thymus was studied by MRI or CT. Biochemical parameters including HSA, globulin, total bilirubin, indirect bilirubin, uric acid, creatinine, and high-density lipoprotein cholesterol (HDL-C) were determined. Individuals were classified according to HSA levels, as "Normal albumin" or "Hypoalbuminemia", defined as HSA ≥ 40g/L and HSA < 40g/L, respectively.

Investigators demonstrated that patients with lower HSA had higher MG severity and a higher rate of myasthenic crises (P < 0.05). Multivariate analysis confirmed these data. The relation between HSA and MG severity remained significant even after adjustment for age, sex, BMI, duration of disease, diabetes, hypertension, cardiopulmonary disease, globulin, WBC, total bilirubin, indirect bilirubin, uric acid, creatinine, HDL-C, and thymus features. Moreover, HSA was positively correlated with uric acid (r = 0.421, P < 0.001), which was inversely correlated with disease severity.

According to this study, HSA is a potential biomarker of MG severity but this conclusion needs to be confirmed also by other research groups. It was uncertain whether treating hypoalbuminemia will improve the outcome of the disease.

The same research group demonstrated with multivariate logistic regression analysis that the albumin to globulin ratio is related to MG severity and prognosis: the lowest values were predictive of longer hospital stays [491].

Uric acid (UA), the final product of purine metabolism, is another potent antioxidant in serum [492, 493]. This can scavenge nitrogen radicals and superoxide, preventing the production of peroxynitrite that can damage cellular metabolism and lipids, carbohydrates, protein, and DNA [494].

Reduced serum UA levels in inflammatory and autoimmune diseases have been detected [495-497]. In the pathogenesis of multiple sclerosis (MS), the UA role has been examined in depth: it can be a marker of disease activity and is one of the antioxidants evaluated for their effect on disease progression [498, 499]. Studies on experimental allergic encephalomyelitis and multiple sclerosis (MS) patients have also indicated a potential therapeutic role of UA in these contexts [500-502].

A study [503] investigated serum UA from 135 patients with MG, 47 patients with MS, and 156 healthy controls (HC). The severity of the disease was established according to the Myasthenia Gravis Foundation of America (MGFA) clinical classification.

Serum UA levels were significantly lower in MG patients than in healthy controls (283 ± 90 vs. 335 ± 84 µM; P < 0.001). The two groups of

MG and MS patients did not differ from each other (283 ± 90 vs. 257 ± 85 µM; P = 0.072). There are not significant differences between early and late onset MG or with/without thymoma or among patients with different duration of disease or depending on bulbar involvement. The relative decrease in the mean serum UA level of the patients with MG correlated inversely with the degree of disease progression, as expressed by the MGFA.

Another study of the same research group also evaluated other antioxidants, i.e., direct, indirect and total bilirubin, albumin and creatinine with UA in 166 MG patients and 214 healthy controls [78]. Significantly reduced levels of all these antioxidants were demonstrated in MG; the relative decrease in the mean serum UA, albumin and creatinine levels of the patients with MG correlated with the degree of disease progression as established by MGFA. Moreover, levels of these antioxidants were lower in female patients in comparison with males. The authors also proposed an integrative therapy for MG patients.

Albumin and prealbumin with several other immune, nutritional and inflammatory biomarkers have also been studied in thymoma MG patients, pre and post thymectomy, with or without preoperative immunonutrition support (i.e., a diet enriched of glutamine, arginine, and polyunsaturated fatty acids, among others, in comparison with a standard nutrition) [504]. In patients with immunonutrition, after thymectomy, postoperative complications were reduced and increased levels of IgA, IgG, IgM, CD3 + T, CD4+T, CD4 + T/CD8 + T, WBC, CRP, and NK-cell were observed, while prealbumin and albumin concentrations were decreased.

In conclusion, in MG there is a low antioxidant status and NMJ can be a target of oxidative stress. It is unknown if the low antioxidant status of serum UA, albumin and other human naturally-occurring antioxidants is a cause or a consequence of MG activity [78]. In the group of naturally occurring antioxidants, albumin and uric acid are the more substantiated biomarkers of activity and disability in MG.

# Chapter 14

# EPIGENETIC MARKERS

Epigenetics concerns heritable alterations in gene expression not modifying the genomic DNA sequence. The environment can modify epigenetic patterning [505]. Currently, epigenetics studies DNA methylation, histone modifications and mainly microRNAs.

A very recent study compared transcriptome and methylome of peripheral monocytes between monozygotic (MZ) twins discordant and concordant for MG, also analyzing MG singletons and healthy controls, all female [506]. This is the first study to define the DNA methylation profile in MG. Results point to an impaired monocyte function in MG and to a decreased expression of genes associated with inflammation resolution that could contribute to the chronicity of the disease. Authors suggested that potential new predictive biomarkers for disease and disease activity could be derived from this line of research.

MicroRNAs (miRNAs) are small (18-25 nucleotides in length) non-coding RNAs found in human and animals, plants and some viruses, that impede gene expression by degrading or blocking translation of their target messenger RNAs (mRNAs) (they function via base-paring to mRNA molecules) and thus regulate essential gene expression patterns [507, 508].

Micro RNAs can be detected in cells or in hematic circulation (but also in other bio-fluids such as urine and saliva) where they are packed into the exosomes or associated with protein complexes such as Argonaute 2 or with HDL particles, and thus are protected from degradation [509]. The real physiological significance of extracellular miRNAs is at present an object of debate, if these molecules mediate cell-to-cell communication or are simply by-products of physiological cell activity [510]. Levels of extracellular miRNAs in the bloodstream have been determined to follow the course of cancer and several autoimmune diseases as systemic lupus erythematosus and multiple sclerosis (MS) appearing as potential biomarkers of disease [511-514].

The first study on miRNAs in MG has been conducted by a research group in Shanghai [515] that revealed let-7 family to be decreased in peripheral blood mononuclear cells (PBMCs) in patients compared to the healthy controls. Interleukin-10 levels could be influenced by this dysregulation and thus be implicated in the pathogenesis of MG.

A subsequent study of the same group [516] showed that miR-320a was downregulated in the PBMCs of MG patients and pro-inflammatory cytokines (among others, IL-2, IFNγ, IL-17) were overexpressed. This dysregulation was derived from the lack of an inhibitory effect of miR-320a on MAPK1 (mitogen-activated protein kinase 1 also known as extracellular signal-regulated kinase 2, ERK2) expression. The authors concluded that miR-320a could be implicated in the modulation of cytokines and the pathogenesis of MG.

Another study [517] evidenced an overexpression of miRNA-146a in PBMCs from MG patients. This has been reported also in rheumatoid arthritis, Sjögren's syndrome and multiple sclerosis patients but not in SLE in which low levels have been described. In experimental autoimmune MG, high levels of miR-146a were also found. This miRNA could interfere with CD40, CD80, Toll like receptor 4 (TLR4) and nuclear factor of kappa-light-chain-enhancer of activated B cells (NF-kB) on AChR specific B cells with a possible role in MG pathogenesis.

Also, miR145 was found to be down regulated in PBMCs and T effector cells of rats with EAMG and MG patients [518]. This miRNA could influence levels of T helper 17 cells and T cell proliferation.

Overexpression of miR145 ameliorated EAMG, raising new possibilities for therapy.

Upregulation of miR-155 was found in peripheral blood mononuclear cells from myasthenia gravis patients [519]. Inhibition of miR-155 reduced AChRAb production in EAMG mice and ameliorated the disease.

A recent study considered the transcriptional profile of peripheral blood mononuclear cells of AChRAb MG with early onset of disease finding that miR-612, miR-3654, and miR-3651 were significantly upregulated in comparison with healthy controls [520].

Other studies considered circulating miRNAs. A research group from Spain investigated the miRNA profile in 61 AChRAb+ MG patients, early onset, late onset and with thymoma, revealing down-regulated miRNAs including miR-15b, miR-122, miR-140-3p, miR-185, miR-192, miR-20b and miR-885-5p [521] without finding increased levels of miR-150-5p and miR-21-5p. Thymoma and treatment did not modify miRNA profile. Another research group [428] reported decreased serum levels of miR-20b in MG patients underlining a negative correlation with QMGSs both in the pretreatment and post treatment stages. Moreover, miR-20b was negatively correlated with IL-8 and IL-25 that are found to be elevated in MG. Levels of miR-20b recovered, and both the interleukins decreased after prednisone acetate.

A study revealed that 2 miRNAs, miR-150-5p that induces T-cell maturation, and miR-21-5p, a regulator of Th1 versus Th2 cell responses, were increased in serum of AChRAb+ female early-onset MG patients [522], whereas miR-27a-3p, involved in natural killer (NK) cell cytotoxicity, was decreased. The miR-150-5p was the most reliable marker showing strict association with MG status and decreasing after thymectomy. Another potential marker was miR27a-3p that seemed to be decreased in MG. Authors concluded that these miRNAs could be biomarkers of MG and in particular miR-150-5p was a marker of the severity of disease.

These data have been considered important findings in the context of considering the role of miR-150-5p in T cell differentiation [523, 524] and the dysregulation of T-cells that seems to be involved in the pathogenesis of MG [390]. The miR-21-5p has been implicated in lupus erythematosus, multiple sclerosis and diabetes type 1 [525-527]; perhaps certain autoimmune diseases share a miRNA profile: miR-21 is induced upon T cell activation and coregulates several aspects of this same process, influencing T cell apoptosis, proliferation and migration [79, 528, 529]. The other potential marker was miR27a-3p which in literature was proposed to reflect natural killer cell cytotoxicity [530].

A subsequent study [79] confirmed and analyzed the usefulness of miR-150-5p and miR-21-5p in MG male and female patients with early and late onset forms of disease, both AChRAb seropositive and seronegative: these molecules resulted in increases in 71 MG patients in comparison with 55 healthy controls (p < 0.0001); after immune-suppressive therapy, were decreased, especially miR-150-5p; in a cohort of patients with psoriasis, Crohn's and Addison's disease, that share the T cell mediated pathophysiology of MG although affecting different organs, miR-150-5p was not significantly different than in healthy controls. The miR27a-3p was not significantly different in comparison with healthy controls and was not considered to be specifically involved in MG. A recent study from the same group found a deregulation of miR-150-5p and miR-21-5p in MG patients after three months of physical training correlating with the amelioration in muscular performances [531]. These results are different from the above mentioned data of researchers from Spain [521]: at present, the reasons for these discrepancies are not clear but perhaps different methodologies can be responsible.

The more rare form of MG with MuSKAb has been considered in one study [80]. Twenty five MuSKAb+ MG patients with heterogeneity in stages of disease and treatment, were considered. Four miRNAs, let-7a-5p, let-7f-5p, miR-151a-3p and miR-423-5p, were significantly increased in serum of MuSKAb+ MG patients. In the preceding study on AChRAb+ MG, these miRNAs were not found to be altered [522]. A preceding study [515] evidenced reduced levels of let-7 miRNAs in peripheral blood mononuclear cells in MG but the antibody status of patients was not reported, impeding comparisons. Members of the let-7 miRNA family are increased in MS patients, in particular let-7a in the secondary progressive form of disease [532]. In literature, let-7 miRNAs are also related to T cell activation through stimulation of the toll-like receptor 7 [533] that in CD4+ T cells induces unresponsiveness [534]. As regards miR-151a-3p (known marker for breast cancer) and miR-423-5p (marker for heart failure), at present there are no clues about correlations with MG pathogenesis. Authors underlined the long duration of MG, immunosuppressive therapy and thymectomy in some cases as confounding factors in their cohort. The signaled miRNAs need to be confirmed in a larger cohort.

Considering the above mentioned data, miRNAs in PBMCs and in circulation are very attractive clues in the pathogenesis of MG. Much research has been conducted on thymic tissue, for example the study of Wang et al. about miRNA 19b [401] (see chapter on markers of T cell, Treg). However, their use as clinical biomarkers is at present very distant.

A very recent area of research in MG concerns the long non-coding RNAs (lncRNA), non-protein coding transcripts longer than 200 nucleotides. These types of RNA influence gene transcription, post-transcriptional regulation and epigenetic regulation and have been found to be involved in several autoimmune and neurological diseases. A single study considered the lncRNA and mRNA profile in MG patients with or without thymoma compared with healthy subjects [535]. A subset of aberrant lncRNAs and mRNAs, potential biomarkers of MG and thymoma, was also described. Using Gene Ontology (GO) and the Kyoto Encyclopedia of Genes and Genomes (KEGG) biological pathway, the authors could infer functions of these RNAs. An analysis of the relationships between lncRNAs and target genes was performed with the 'cis' and 'trans' model. A series of differentially expressed genes, in particular regarding cytokines, was so reported in MG patients with or without thymoma, giving insights into the pathogenesis of disease.

# Chapter 15

# GENETIC DNA MARKERS

Autoimmune myasthenia gravis is determined by environmental and genetic factors. In a twin study, 11 of 31 pairs (35.5%) of monozygotic twins were concordant for MG; none of 16 dizygotic twins were concordant [536]. However the true DZ rate may be similar to the familial rate, between 3.8 and 7.2% [537, 538] whereas the worldwide frequency is approximately 1 in 10,000 individuals. Twin studies show a heritability index of 0.65, a level of genetic predisposition comparable to Alzheimer's disease and epilepsy and above multiple sclerosis [177]. Apparently, these data underscore environmental factors in MG etiology recognizing the importance of genetic background. From the genetic point of view, MG is a complex disease with numerous genetic polymorphisms, each giving a small contribution to MG predisposition. Familial cases are rare, representing 3.8–7.2% [538, 539]. An Italian-American family with five siblings affected by late onset MG has been recently described with a sequence variant in the ecto-NADH oxidase 1 gene (ENOX1) [540]. This variant sequence was not found in sporadic or familial autoimmune MG. The ENOX1 variant decreased significantly the ENOX1 expression in a gene dose-dependent manner. Mechanisms of the ENOX1 determined predisposition are unknown. Other autoimmune disorders can be found in families with MG patients suggesting a common genetic and environmental predisposition to autoimmunity [541, 542]. Genetic research evidenced specific HLA alleles involved in MG susceptibility. The HLA A1-B8-DR3-DQ2 haplotype, also known as AH8.1, is associated with EOMG in the Caucasian population [207, 543-545]. This haplotype is also associated with other autoimmune diseases suggesting common predisposing genetic factors [546, 547]. Other HLA associations have been described with geographic restriction [505]. In Chinese southern Han, a particular group of childhood-onset ocular MG has been reported to be positively associated with DQ9 haplotype (in 90.1% of patients) [548]. In Chinese northern Han MG patients, HLA-DRB1(*)09 allele was found to be significantly represented [549].

In Norway and Italy late onset MG has been associated with DRB1*15:01, DQB1*05:02 and DRB1*16 [550, 551]. In Italy, the frequency of the TNF-B*1, C4A*Q0, C4B*1, DRB1*03 supratype was found to be increased in female patients with early onset MG [552].

A very recent study in a Portuguese cohort [553] confirmed the association with the HLADRB1*03 and HLA B*08 in the global MG population and in the EOMG subgroup, and signaled a new association of LOMG with the HLA DRB1*01 allele. A thymoma MG subgroup was associated with the HLA DRB1*10 allele. The authors observed that preceding studies about thymoma MG patients found an association with HLA DRB1*11 in Mexican mestizo patients [554] and with HLA DQA1*0401 and HLADQB1*0604 in Chinese patients [555]. Regarding differences of association according to histological subtypes of thymoma, a study on 78 Caucasian French MG patients found a protective effect of HLA-A02 for B2 type thymoma and an increase of frequency of HLA-A25 in the whole group of patients [556]. A study signaled that both thymomatous MG and titin positive nonthymomatous MG are associated with TNFα, TNFβ, Fcγ receptor IIa and IL-10 genotypes with a risk increasing with the number of such genetic markers and with a prevalent contribute of Fcγ receptor IIa allelic variants for thymoma MG [557].

The MG form with MuSKAb has been associated with DQ5 alleles in southern and northern Europe [81, 82].

A recent genome-wide association study (GWAS) from northern Europe (649 patients) signaled associations of early onset AChRAb+ MG with the HLA class 1 region (specifically HLA-B*08), with the 'Protein Tyrosine Phosphatase, Non-Receptor Type 22' (PTPN22) gene and with the 'TNFAIP3-interacting protein 1' (TNIP1) gene (a new association) [83].

The PTPN22 (also called lymphoid-specific phosphatase) is bound to C-src tyrosine kinase in cells (Csk) [558]; the complex of these two proteins inhibits T-cell-receptor signaling [559]. The change of amino acid at position 620 from an arginine (R) to a tryptophan (W) in PTPN22 disrupts the interaction between PTPN22 and Csk, impeding the formation of the complex and the suppression of T-cell activation. The T-allele of PTPN22 binds less efficiently to Csk than the C-allele does, consequently T-cells expressing the T-allele may be hyper-responsive and individuals carrying this allele may develop autoimmunity [558-561]. The W620 variant has been shown to be significantly associated with MG patients (31% of EOMG versus 20% of healthy controls) and with higher titers of AChRAbs [505, 562]. An association of PTPN22gain-of-function+1858T(+) genotypes with EOMG and thymoma MG has been reported [563], with a low expression of IL-2 in thymomas that could be implicated in the MG pathogenesis in this subgroup of patients.

TNIP1 is a risk allele in other autoimmune and inflammatory diseases [505, 564-566] and is implicated in ubiquitin-dependent dysregulation of NF-kB signaling [567, 568].

Another GWAS from North America (1032 patients) revealed both early and late onset AChRAb+ MG groups associations with HLA class 2 locus (HLA-DQA1 but with different sets of single-nucleotide polymorphisms within the two groups), with the 'cytotoxic T-lymphocyte–associated protein 4' gene (CTLA4) and, for the late onset MG group, with 'tumour necrosis factor receptor 4 superfamily, member 11a, NF-κB activator' (TNFRSF11A) [84, 569]. These findings substantially replicated for CTLA4 and HLA-DQA1 in an independent cohort of Italian MG patients in the same study. A predisposing effect of some alleles of CTLA4 gene to general risk of MG was demonstrated in the northern Chinese population [570].

CTLA4 is a 45-kD immunoglobulin protein expressed by activated T cells. This protein increases T-cell motility and reduces contact periods between T cells and antigen-presenting cells consequently decreasing cytokine production and cell proliferation [84]. In this manner, CTLA4 would downregulate T-cell activation, terminate T-cell responses, and protect against autoimmunity [84, 571]. These functions of CTLA4 were substantially confirmed by CTLA4-deficient mice [572]. As regards mechanisms of action of CTLA4, cells expressing CTLA4 avidly endocytose the costimulatory factors CD80 and CD86 on antigen-presenting cells and degrade them [84, 573]. The CTLA4 locus has been implicated in other autoimmune disorders [574-577]. Proteins consisting of CTLA4 rendered soluble by fusion to antibodies (CTLA4-IgG) have been used in rats with EAMG [578, 579]. Certain SNPs in the promoter region of CTLA4 have been found to be associated with MG [580-582]. Perhaps these SNPs might cause aberrant splicing of CTLA4 and/or cellular abnormalities of T-cells contributing to the disease pathogenesis. In fact, a particular spectrum of CTLA4 mRNA expression due to alternative splicing in MG patients has been described [582]. The CTLA4(low) +49G(+) genotypes were shown to be more frequent in late onset MG (age at onset $\geq 60$ years) as compared with controls [583]. A preceding study had shown no statistically significant association of the +49 G/G genotype with myasthenia gravis in comparison with healthy controls [584]. Earlier studies reported other polymorphisms of CTLA4 gene in particular groups of MG patients in association with genetic variations of the interleukin-1 gene [585, 586].

The gene of TNFRSF11A encodes a 4.5-kDa receptor activator of nuclear factor-κ B expressed on the surface of antigen-presenting dendritic cells [587]. This receptor regulates the interaction between T cells and dendritic cells and is crucial for lymph node organogenesis and osteoclast differentiation [588].

The gene of TNF-α is located within the HLA locus tightly linked to AH8.1 haplotype. The TNF-α is a pro-inflammatory cytokine implicated in autoimmune and inflammatory conditions. The importance of this locus has been demonstrated by several studies. A functional SNP, rs1800629, at −308 nucleotides upstream from the transcription initiation site can affect TNF-α expression, with the −308A allelic form having a two-fold greater level of transcription than the 308G form [589]. This allelic form was related to elevated serum levels of TNF-α and with a severe disease course in several diseases [590-593]. The A allele of this functional SNP has been correlated to increased levels of TNF-α in MG [594-596]. Another SNP, rs1800629, is significantly associated with EOMG mainly in females [505].

The specific transcription factor of Treg, fork head/winged-helix transcription factor (FOXP3), regulates the number and function of Tregs [597]. Both the number and function of Treg are found by some authors to be impaired in MG (see the chapter on T cell markers) [390]. A study found a negative relation between an allele of FOXP3 (FOXP3 IVS9+459Gallele) and MG in a Han Chinese population [598]. Polymorphisms in CHRNA1 and CHRND are probably minor susceptibility genes for developing MG [599-601]. A SNP (SNP, rs16862847) in the promoter region of CHRNA1 (gene coding for the alpha subunit of the AChR) has been signaled that resulted more frequently in MG patients in two different cohorts from France and UK [602].

Gene TBX21 encodes the transcription factor T-bet that prevents Th1 to Th2 cell differentiation [603]. The importance of T-bet in autoimmunity is easily deducible and has been underlined by recent advances on inflammatory bowel diseases [604]. Nevertheless, there are very few data on MG. Mice with T-bet deficiency, T-bet(-/-) mice, are less susceptible to induction of experimental MG and characterized by a reduction of autoreactive Th1 cells and an increase of Th2, Th17 and Foxp3+ Treg cells [605]. A role of genetic variants regarding T-bet in MG is probable but has not been explored so far.

The cysteine protease cathepsin V (gene CTSL2) is involved in antigen presentation in thymic epithelial cells. A polymorphism of this gene has been reported to be associated with early-onset MG (p = 0.03) [606].

A study evaluated genetic polymorphisms of two regulators of T-cell activation, Galectin-1 (LGALS1) and interleukin receptor 2β (IL2Rβ) in 146 Caucasian myasthenia gravis patients found a significant association for the rs4820293/rs743777 polymorphism haplotypes (p < 0.01). The interaction between LGALS1 and IL2Rβ could be involved in MG pathogenesis [607].

A polymorphism of interferon γ, +874T, has been reported to be less frequent in MG but it did not remain significant after correction for multiple comparisons [608].

As regards the IL-4 gene, the GG rare homozygote genotype of the I75V polymorphism has been shown to be associated with myasthenia gravis. Perhaps a reduced responsiveness to interleukin-4 might be involved in disease pathogenesis [609].

A polymorphism (the ACC/ACC haplotype) of the gene of IL-10 was found to be associated with MG and particularly with the subgroups with late onset MG, thymomatous and anti Titin positive MG. Early onset MG was associated with the ATA/ATA haplotype [610]. Successively, a statistical trend of association with IL-10 promoter SNPs and early and late-onset MG was revealed by Zagoriti et al. [611].

IgG receptors (FcgammaR) are important in immune response and inflammation. FcγRIII has been demonstrated to be involved in the pathogenesis of experimental MG in mice [612]. Functional poly-morphisms of FcγR genes affect the efficiency of FcgammaR-mediated functions. A polymorphism of FcgammaRIIa, R/R131 genotype, has been shown to be associated with MG in Dutch patients and has been qualified as a marker for MG susceptibility [613].

A genetic study with candidate gene approach in early onset MG found a significant role for some novel loci: CD86 (CD28 antigen ligand 2, B7-2 antigen), AKAP12 (A kinase anchor protein 12), VAV1 (vav 1 guanine nucleotide exchange factor, critical regulator of T-cell cycle progression and proliferation) and BAFF (see chapter on inflammatory proteins) [614]. The more consistent genes VAV1 and BAFF interacted epistatically with a greater risk of the two in combination. These two genes and CD86 share the NFkB signaling pathway.

A recent study on African juvenile MG patients who develop severe/complete treatment-resistant ophthalmoplegia showed fourfold higher odds of harboring a functional transforming growth factor beta-1, TGFB1 c. − 387 T containing haplotype, in comparison with racially appropriate MG controls without this complication [615]. This mutation provokes lower basal TGFB1, inhibiting upregulation of extraocular muscle DAF/CD55 (decay accelerating factor, a complement regulator that defends against complement-mediated damage in NMJ of MG) and/or altering endplate remodeling. The same group previously described the c.-198C > G DAF polymorphism implicating a greater risk of ophthalmo-plegia in MG (this mutation would prevent DAF upregulation) [616] and specified that MG patients with this mutation may be at higher risk because of prednisone treatment [617]. CD59 is another complement regulator but mutations of its gene have not been described in human MG. The importance of CD55 and CD59 in protection against complement mediated injury has been previously demonstrated [618].

A whole genome-based single nucleotide polymorphisms analysis was conducted on 109 MG patients and 150 controls without neurological disease [619]. Among other genetic polymorphisms, RYR3 (ryanodine receptor), CACNA1S (dihydropyridine receptor), and SLAMF1 (an important genetic susceptibility locus for autoantibody production) were reported but the authors underlined the need for confirmation with larger samples.

Recently, RNA sequencing has been used to evaluate gene expression in blood in MG patients [620]. The transcriptome was studied in 8 patients with MG who were either in remission or active disease states. Twenty-eight genes that are differentially expressed according to disease activity have been reported in this small cohort of patients. Functional analysis of these genes showed that they are involved in immune cell trafficking (ICAM1, CCL3, S100P and GAB2) and apoptosis (NFKBIA, ZC3H12A, TNFAIP3, and PPP1R15A) and could be biomarkers of MG activity.

Pharmacogenetics studies inherited genetic differences in drug metabolic pathways determining individual responses to drugs to predict responsiveness and adverse effects.

As regards MG, some studies considered pharmacogenetics of glucocorticoids (GC) in this group of patients. The rs17209237*G allele in the GC receptor gene was found to be associated with GC responsiveness (probable marker for GC sensitivity in MG) [621]. The same research team subsequently reported the association of rs11728697 SNP, and of a risk haplotype (AGTACT) containing a mutation at this SNP, in the secreted phosphoprotein 1 (*SPP1*) gene (coding osteopontin, a pro-inflammatory cytokine), with a poor response to 3-month GC treatment [622]. In PBMC of MG patients, the P-glycoprotein (a drug efflux pump that transport drugs out of target cells such as lymphocytes decreasing their efficacy) function is increased by long term therapy or high dosages of GC [623] and decreased by tacrolimus [624]. High-dose administration of GCs results in increased expression of mRNA and related protein of P-glycoprotein [624].

Pharmacogenetic profiling of azathioprine is currently centered on analysis of gene coding for thiopurine S-methyltransferase (*TPMT* gene), some mutations of which result in unresponsiveness and others in increased side effects due to low enzymatic activity and accumulation of cytotoxic 6-thioguanine nucleotides. Hyperactivity of TPMT decreases pharmacologic activity. A deficit of TPMT (3-6 out of 1000) or reduced TPMT activity (approximately 11%) significantly increases the risk of drug-induced leukopenia and myelosuppression, bleeding, infection and even death [625]. These mutations are markers of toxicity for azathioprine use [177] and TPMT genotyping (and dosing determination based on TPMT genotype) is recommended before azathioprine use [626]. Sometimes a toxicity marker could not associate with adverse effects as reported by Colleoni et al. for TPMT *3A haplotype [627]. Consequently these authors underline that pharmacogenetic profiling for azathioprine should also include other genes of metabolism of azathioprine such as glutathione-S-transferase (see also [627] and [628]), xanthine oxidase and aldehyde oxidase.

Pharmacogenetics of methotrexate (MTX) have been extensively studied in literature, particularly in rheumatoid arthritis (RA). A recent metanalysis signaled that the RFC-1 80G > A (rs1051266) polymorphism

(MTX transporter influencing cellular uptake and efflux) is associated with MTX toxicity in European patients with RA [629]. In MG there are insufficient data. The polyglutamated form of MTX in RBC was used in the recent trial of methotrexate in MG as a secondary outcome measure because it could be representative of intracellular MTX levels in target tissues and may potentially predict a response to the drug (630]. This biomarker has been used for the effectiveness of MTX in RA [631, 632]. The polyglutamated form of MTX is derived from the action of folylpolyglutamyl synthase (FPGS). Upregulation of FPGS and downregulation of ATP-binding cassette G2 (ABCG2) that promotes MTX efflux, would characterize the response to folate deprivation realizing cellular retention of folates. Allelic variation in these genes could result in an increased or decreased polyglutamated form of MTX and could influence response to MTX. Another recent metanalysis on RA [633] found that the absence of TYMS 1494 del6 and FPGS rs10106 and presence of MTHFR C677T predict adverse events with MTX.

Cyclosporine is a substrate for intestinal P-glycoprotein encoded by the multidrug resistance 1 (ABCB1) gene. A study on pharmacogenetics of cyclosporine in MG patients evidenced that ABCB1 polymorphisms in both genotype and haplotype may have a minor effect on its blood concentrations [634].

Although these studies are attractive, the direct relevance of pharmacogenomics markers to the toxic and/or therapeutic effects of drugs has not been yet established [635].

As regards the use of intravenous immunoglobulin in MG, 63 affected patients were studied to search differences in response to treatment with polymorphisms within the FCγR2A (rs1801274), FCγR2B (rs1050501), FCγR3A (rs396991), and FCγR3B (NA1/NA2) genes [636]. No significant difference was found in any of the polymorphisms studied between responders and nonresponders, but the FCγR2B-232I/I polymorphism was shown to be related to higher disease severity as measured with QMGS.

*Chapter 16*

# SEX HORMONES

Autoimmune diseases are more prevalent in women than in men [505]. In several autoimmune conditions, androgens are considered to be protective and estrogens to mediate autoimmunity. In MG, women show a higher incidence and earlier age of onset compared to men, and disease fluctuations in pregnancy are well described [637]. In a study [638], an increased expression of α estrogen receptor was demonstrated on thymocytes and both α and β receptors on T cells from peripheral blood mononuclear cells in MG patients. Moreover, pro-inflammatory cytokines could induce increased expression of estrogen receptors in normal thymocytes mainly in the CD4+ subset. Another study [639] demonstrated in EAMG that treatment with 17β-estradiol before Ag exposure is necessary and sufficient to promote AChR-specific Th1 cell expansion. This pretreatment with 17β-estradiol has been shown to increase the severity of EAMG in mice. These data and other evidence in literature demonstrate that estrogens have an important role in experimental and human MG.

Other authors [640] have evidenced the role in early-onset MG of several genes (HLA-G, TAP2, HLA-DRB1, TUBB) under the potential control of sex hormones, localized in the MHC and surrounding region.

These genes contain estrogen response elements and their polymorphisms might influence transcription in a gender specific manner.

In early onset MG, a gender bias has been reported from the initial descriptions. A more recent study [641] analyzed a series of HLA region markers in 207 Caucasoid patients affected by MG with onset before age 40 and the associations with HLA-DR3 and -B8 were significantly stronger in the 165 females than in the 42 males. A GWAS [83] has confirmed a strong difference in risk for HLA-B8 when comparing female and male patients with MG. More precisely, HLA-B8 was more associated with EOMG in females (OR, 6.92; 95% CI, 4.28–5.91) than in males (OR, 3.55; 95% CI, 2.46–4.56): a significant difference between the 2 groups (p < 0.0001) was found. The authors pointed out that this sex bias may reflect hormonal effects, a female bias in environmental exposure, or gender-related epigenetic changes playing a role in disease risk.

*Chapter 17*

# CONCLUSION

Different forms of MG are defined according to autoimmune and antibody disease mechanisms, target molecules of skeletal muscle, thymic status, genetic characteristics, response to therapy, and disease phenotype [642]. Each patient should be assigned to a specific subgroup and it implicates different therapeutic decisions and prognosis.

In ocular MG (15% of MG cases), half of patients have AChRAb seropositivity, a minority can have LRP4Ab seropositivity in isolation or in addition to AChRAb. The RIA seropositivity for MuSKAbs is exceptional. Ninety percent of patients who have only ocular symptoms and signs for 2 years will continue to have an ocular form of the disease [642]. Immunosuppression can be considered in several cases, thymectomy is not recommended.

Two thirds of MG patients have generalized early (<50 years, more commonly female) or late (≥50 years, slight male prevalence) onset MG, with a high frequency of AChRAb seropositivity and the same generalized clinical phenotype, without thymoma [642]. Juvenile MG, defined by an onset before the age of 15 years, has the same course of early onset MG and is more common in East Asian populations. Early onset MG is characterized by thymic hyperplasia, is frequently associated with other immune diseases such as thyroiditis (in 15% of cases) and is often HLA

B8DR3 positive or positive for other non-HLA immune related genes. The late onset form is characterized by thymic atrophy and frequent positivity for HLA B7 DR2 haplotype and HLA DR B1 15.01 [643]. Early and especially late onset MG often necessitates immunosuppressive therapy. Thymectomy is indicated in early onset form. In late onset MG, thymectomy seems to be useless excepting a subgroup of patients with onset age at 50-65 years and thymic hyperplasia, who could benefit from it.

Ten percent of MG patients have a thymoma (paraneoplastic MG) and probably a different HLA genotyping. Antibodies against AChRs are nearly always present. Thymectomy is necessary, immunosuppression is very frequently required.

The presence of AChRAbs is a good biomarker for the diagnosis of MG. Antibody titers are not reliable as a severity index in the comparison between different patients, although in the same patient higher levels can correlate with relapses.

Antibodies against MuSK are a biomarker for a particular form of MG highly prevalent in females, sometimes with muscular atrophy, without thymic disease. These patients have a frequent impairment of bulbar muscles and respiratory crises. They can present also with the involvement of neck, shoulder, and respiratory muscles. Several patients are indistinguishable from generalized AChRAb+ MG. A less favorable response to anticholinesterases has been reported in this form of MG. Immunosuppressive therapy is frequently required and rituximab seems to be particularly effective. Thymectomy is generally useless. Following a patient over time, MuSKAb titers tend to correlate with the course (severity and activity) of disease [104].

LRP4 antibodies are rare (1 to 5% of all MG patients), typical of mild/moderate disease, without thymic disease. Thymectomy is not recommended. These antibodies can be present in association with MuSKAbs or AChRAbs and in these cases the general indications for these latter forms of MG seem to prevail. Also, they can be found in other autoimmune disorders and ALS with uncertain significance [104].

Seronegative MG patients (about 10% of generalized MG patients) should be re-examined with RIPA assay 6-18 months after the initial assessment [104]. If tested with a cell-based assay, one third of RIPA seronegative generalized MG patients will result AChRAb, MuSKAb or LRP4Ab seropositive. In proven seronegative patients, the distinction between early and late onset does not seem to have clinical or prognostic relevance.

As regards striational antibodies, they are not specific for MG and are found also in several autoimmune conditions and in thymoma without MG. In addition, striational antibodies are found almost always in association with AChRAbs and thus their MG diagnostic value is scarce. They can be good markers for thymoma in early onset AChRAb+ MG. Titin and ryanodine receptor antibodies are reported in association with thymoma and in late onset MG and can be markers for severe diseases. In general, striational antibodies' seropositivity is an indication for immunosuppression [642]. Titin and ryanodine receptor antibodies in late onset MG could be indicative of no response to thymectomy but controlled studies are necessary to confirm it.

Kv1.4 antibodies have been associated with severe MG and heart disease in Japanese patients but this does not seem to be confirmed in European ancestry.

Agrin, cortactin, rapsin and collagen Q antibodies do not have a definite diagnostic/prognostic marker value at present. Agrin antibodies do not seem to be specific to MG because they are been detected in other diseases such as ALS.

As regards the other potential markers of MG, evidence from literature is still insufficient for their use in clinical practice.

Lower frequencies or defective regulatory B and T cells, higher frequencies of plasma cells, Th17 and follicular CD4+ Th cells, respective cytokines and surface markers, are all described in MG but would require an accurate cut-off of determination, statistically representative sample of patients and replication of the data from different research groups, possibly without gender or racial bias. Moreover, a clear added value to current recognized serum biomarkers (antibodies) is necessary in the clinical context. The same can be said for other inflammation-related molecules and the reported naturally occurring antioxidants.

MicroRNAs apparently have good potential as humoral markers for MG because of their relative abundance, highly specific expression, and stable presence in serum and other biofluids [177]. Nevertheless, different groups of researchers reported nonoverlapping sets of miRNAs for the same disease because of different methodologies. Recently, it has been recommended to avoid [510] contamination of the samples from intact cells and to utilize optimal methods for miRNA extraction (Exiqon miRCURY TM Biofluids Kit for plasmatic samples). The method of quantification of circulating miRNAs is quantitative real-time reverse transcription polymerase chain reaction (qRT-PCR) but the problem of normalization of miRNA expression data has still not been resolved yet. Moreover, only clinically homogeneous groups of patients should be compared, for example with the same MG forms or with/without therapy. All of these are the present obstacles to the practical application of miRNAs as biomarkers of MG.

Very few genetic markers have entered into clinical practice, essentially HLA markers that are of recognized epidemiological value. Pharmacogenetic profiling is recommended in azathioprine therapy to prevent adverse effects. Moreover, a TGFB1 haplotype and a DAF polymorphism related to severe treatment resistant ophthalmoplegia in MG have been reported. However, with these exceptions, the practical clinical significance of the other genetic data discussed in this chapter remains to be defined.

# REFERENCES

[1] Carr, A. S., Cardwell, C. R., McCarron, P. O. and McConville, J. (2010). A systematic review of population based epidemiological studies in myasthenia gravis. *BMC Neurol*, 18 (10):46.

[2] Binks, S., Vincent, A., Palace, J. (2016). Myasthenia gravis: a clinical-immunological update. *J Neurol*, 263 (4): 826-834.

[3] Gwathmey, K. G. and Burns, T. M. (2015). Myasthenia Gravis. *Semin Neurol*, 35 (4): 327-339.

[4] Melzer, N., Ruck, T., Fuhr, P., Gold, R., Hohlfeld, R., Marx, A., Melms, A., Tackenberg, B., Schalke, B., Schneider-Gold, C., Zimprich, F., Meuth, S. G. and Wiendl, H. (2016). Clinical features, pathogenesis, and treatment of myasthenia gravis: a supplement to the Guidelines of the German Neurological Society. *J Neurol*, 263 (8):1473-1494.

[5] Skeie, G. O., Apostolski, S., Evoli, A., Gilhus, N. E., Illa, I., Harms, L., Hilton-Jones, D., Melms, A., Verschuuren, J., Horge, H. W. and European Federation of Neurological Societies. (2010). Guidelines for treatment of autoimmune neuromuscular transmission disorders. *Eur J Neurol*, 17 (7):893-902.

[6] Cooper, M. D. and Alder, M. N. (2006). The evolution of adaptive immune systems. *Cell*, 24; 124 (4): 815-822.

[7]    Holtmeier, W. and Kabelitz, D. (2005). Gammadelta T cells link innate and adaptive immune responses. *Chemical Immunology and Allergy*, 86: 151–183.

[8]    Miller, J. F. (1993). Self-nonself discrimination and tolerance in T and B lymphocytes. *Immunol Res*, 12 (2): 115–130.

[9]    Sproul, T. W., Cheng, P. C., Dykstra, M. L. and Pierce, S. K. (2000). A role for MHC class II antigen processing in B cell development. *Int Rev Immunol*, 19 (2-3): 139-155.

[10]   Zdrojewicz, Z., Pachura, E. and Pachura, P. (2016). The Thymus: A Forgotten, But Very Important Organ. *Adv Clin Exp Med*, 25 (2): 369-375.

[11]   Derbinski, J., Schulte, A., Kyewski, B. and Klein, L. (2011). Promiscuous gene expression in medullary thymic epithelial cells mirrors the peripheral self. *Nat Immunol*, 2: 1032-1039.

[12]   Suniara, R. K., Jenkinson, E. J. and Owen, J. J. (2000). An essential role for thymic mesenchyme in early T cell development. *J Exp Med*, 191: 1051-1056.

[13]   Le Panse, R. and Berrih-Aknin, S. (2005). Thymic myoid cells protect thymocytes from apoptosis and modulate their differentiation: implication of the ERK and Akt signaling pathways. *Cell Death Differ*, 12: 463-472.

[14]   Willcox, N., Leite, M. I., Kadota, Y., Jones, M., Meager, A., Subrahmanyam, P., Dasgupta, B., Morgan, B. P. and Vincent, A. (2008). Autoimmunizing mechanisms in thymoma and thymus. *Ann NY Acad Sci*, 1132: 163–173.

[15]   Berrih-Aknin, S. and Le Panse, R. (2014). Myasthenia gravis: a comprehensive review of immune dysregulation and etiological mechanisms. *J Autoimmun,* 52: 90-100.

[16]   Sleckman, B. P. (2005). Lymphocyte antigen receptor gene assembly: Multiple layers of regulation. *Immunol Res*, 32 (1–3): 253–258.

[17]   Peterson, P. R., Org T. N. and Rebane, A. (2008). Transcriptional regulation by AIRE: Molecular mechanisms of central tolerance. *Nat Rev Immunol*, 8 (12): 948-957.

[18] Raker, V. K., Domogalla, M. P. and Steinbrink K. (2015). Tolerogenic Dendritic Cells for Regulatory T Cell Induction in Man. *Front Immunol*, 9, 6:569.

[19] Macián, F., García-Cózar, F., Im, S. H., Horton H. F., Byrne M. C. and Rao, A. (2002). Transcriptional Mechanisms Underlying Lymphocyte Tolerance. *Cell*, 109 (6): 719–731.

[20] Rudensky, A. Y., Gavin, M. and Zheng Y. (2006). FOXP3 and NFAT: Partners in Tolerance. *Cell*, 126 (2): 253–256.

[21] Soto-Nieves, N., Puga, I., Abe, B. T., Bandyopadhyay, S., Baine, I., Rao, A. and Macián, F. (2009). Transcription complexes formed by NFAT dimers regulate the induction of T cell tolerance. *J. Exp. Med*, 206 (4): 867–876.

[22] Bettelli, E., Carrier, Y., Gao, W., Korn, T., Strom, T. B., Oukka, M., Weiner, H. L. and Kuchroo, V. K. (2006). Reciprocal developmental pathways for the generation of pathogenic effector TH17 and regulatory T cells. *Nature May 2006*, 441 (7090): 235–238.

[23] Wira, C. R., Crane-Godreau, M. and Grant, K. Endocrine regulation of the mucosal immune system in the female reproductive tract. In: Ogra, P. L., Mestek, J., Lamm, M. E., Strober, W., McGee, J. R., Bienestock, *J. Mucosal Immunology.* San Francisco: Elsevier.

[24] Fimmel, S. and Zouboulis, C. C. (2005). Influence of physiological androgen levels on wound healing and immune status in men. *Aging Male*, 8 (3-4): 166–174.

[25] von Essen, M. R., Kongsbak, M., Schjerling, P., Olgaard, K., Odum, N. and Geisler, C. (2010). Vitamin D controls T cell antigen receptor signaling and activation of human T cells. *Nat Immunol*, 11 (4): 344–349.

[26] Sigmundsdottir, H., Pan, J., Debes, G. F., Alt, C., Habtezion, A., Soler, D. and Butcher, E. C. (2007). DCs metabolize sunlight-induced vitamin D3 to 'program' T cell attraction to the epidermal chemokine CCL27. *Nat Immunol*, 8 (3): 285–293.

[27] Dorshkind, K. and Horseman, N. D. (2000). The roles of prolactin, growth hormone, insulin-like growth factor-I, and thyroid hormones in lymphocyte development and function: insights from genetic

models of hormone and hormone receptor deficiency. *Endocr Rev*, 21 (3): 292–312.

[28] Nagpal, S., Na, S. and Rathnachalam, R. (2005). Non-calcemic actions of vitamin D receptor ligands. *Endocrine Reviews*, 26 (5): 662–87.

[29] Drachman DB. (2016). Myasthenia Gravis. *Semin Neurol*, 36(5):419-424.

[30] Cavalcante, P., Galbardi, B., Franzi, S., Marcuzzo, S., Barzago, C., Bonanno, S., Camera, G., Maggi L., Kapetis D., Andreetta F., Biasiucci A., Motta T., Giardina C., Antozzi C., Baggi F., Mantegazza, R. and Bernasconi, P. (2016). Increased expression of Toll-like receptors 7 and 9 in myasthenia gravis thymus characterized by active Epstein-Barr virus infection. *Immunobiology*, 221(4):516-527.

[31] Berrih-Aknin, S. and Le Panse, R. (2014). Myasthenia gravis: A comprehensive review of immune dysregulation and etiological mechanisms. *J Autoimmun*, 52: 90-100.

[32] Berrih-Aknin, S., Morel, E., Raimond, F., Safar, D., Gaud, C., Binet, J. P., Levasseur, P. and Bach, J. F. (1987). The role of the thymus in myasthenia gravis: immunohistological and immunological studies in 115 cases. *Ann N Y Acad Sci*, 505: 50-70.

[33] Keijzers, M., Nogales-Gadea, G. and de Baets, M. (2014). Clinical and scientific aspects of acetylcholine receptor myasthenia gravis. *Curr Opin Neurol*, 27(5): 552-557.

[34] Willcox, N., Leite, M. I., Kadota, Y., Jones, M., Meager, A., Subrahmanyam, P., Dasgupta, B., Morgan, B. P. and Vincent, A. (2008). Autoimmunizing mechanisms in thymoma and thymus. *Ann NY Acad Sci*, 1132: 163–173.

[35] Marx, A., Pfister, F., Schalke, B., Saruhan-Direskeneli, G., Melms, A. and Strobel, P. (2013). The different roles of the thymus in the pathogenesis of the various myasthenia gravis subtypes. *Autoimmun Rev*, 12: 875–884.

[36] Oger, J. and Frykman, H. (2015). An update on laboratory diagnosis in myasthenia gravis. *Clin Chim Acta*, 20; 449:43-48.

[37] Huete-Garin, A., Sagel, S. S. (2005). Chapter 6: "Mediastinum, Thymic Neoplasm". In: J. K. T. Lee; Sagel, S. S., Stanlej, Heiken, J. P. *Computed Body Tomography with MRI Correlation,* Philadelphia: Lippincott Williams & Wilkins. 311-324.

[38] Bernard, C., Frih, H., Pasquet, F., Kerever, S., Jamilloux, Y., Tronc, F., Guibert, B., Isaac, S., Devouassoux, M., Chalabreysse, L., Broussolle, C., Petiot, P., Girard, N. and Sève, P. (2016). Thymoma associated with autoimmune diseases: 85 cases and literature review. *Autoimmunity reviews,* 15 (1): 82–92.

[39] Evoli, C., Minisci, C., Di Schino, C., Marsili, F., Punzi, C., Batocchi, A. P., Tonali, P. A., Doglietto, G. B., Granone, P., Trodella, L., Cassano. A. and Lauriola, L. (2002). Thymoma in patients with MG: characteristics and long-term outcome. *Neurology,* 59 (12): 1844–1850.

[40] Meriggioli, M. N. and Sanders, D. B. (2009). Autoimmune myasthenia gravis: emerging clinical and biological heterogeneity. *Lancet Neurol,* 8(5): 475-490.

[41] Marx, A., Willcox, N., Leite, M. I., Chuang, W. Y., Schalke, B., Nix, W. and Ströbel, P. (2010). Thymoma and paraneoplastic myasthenia gravis. *Autoimmunity,* 43(5-6): 413-427.

[42] Buckley, C., Douek, D., Newsom-Davis, J., Vincent, A. and Willcox, N. (2001) Mature, long-lived CD4+ and CD8+ T cells are generated by the thymoma in myasthenia gravis. *Ann Neurol,* 50: 64–72.

[43] Hoffacker, V., Schultz, A., Tiesinga, J. J., Gold, R., Schalke, B., Nix, W., Kiefer, R., Muller-Hermelink, H. K. and Marx, A. (2000). Thymomas alter the T-cell subset composition in the blood: a potential mechanism for thymoma-associated autoimmune disease. *Blood,* 96: 3872–3879.

[44] Strobel, P., Helmreich, M., Menioudakis, G., Lewin, S. R., Rudiger, T., Bauer, A., Hoffacker, V., Gold, R., Nix, W., Schalke, B., Elert, O., Semik, M., Muller-Hermelink, H. K. and Marx, A. (2002). Paraneoplastic myasthenia gravis correlates with generation of mature naïve CD4(+) T cells in thymomas. *Blood,* 100: 159–166.

[45] Strobel, P., Murumagi, A., Klein, R., Luster, M., Lahti, M., Krohn, K., Schalke, B., Nix, W., Gold, R., Rieckmann, P., Toyka, K., Burek, C., Rosenwald, A., Muller-Hermelink, H. K., Pujoll-Borrell, R., Meager, A., Willcox, N., Peterson, P. and Marx, A. (2007). Deficiency of the autoimmune regulator AIRE in thymomas is insufficient to elicit autoimmune polyendocrinopathy syndrome type 1 (APS-1). *J Pathol*, 211: 563–571.

[46] Strobel, P., Rosenwald, A., Beyersdorf, N., Kerkau, T., Elert, O., Murumagi, A., Sillanpaa, N., Peterson, P., Hummel, V., Rieckmann, P., Burek, C., Schalke, B., Nix, W., Kiefer, R., Muller-Hermelink, H. K. and Marx, A. (2004). Selective loss of regulatory T cells in thymomas. *Ann Neurol*, 56: 901–904.

[47] Marx, A., Willcox, N., Leite, M. I., Chuang, W. Y., Schalke, B., Nix, W. and Strobel, P. (2010). Thymoma and paraneoplastic myasthenia gravis. *Autoimmunity*, 43: 413–427.

[48] Dardenne, M., Savino, W. and Bach, J. F. (1987). Thymomatous epithelial cells and skeletal muscle share a common epitope defined by a monoclonal antibody. *Am J Pathol*, 126: 194–198.

[49] Unusual Cancers of Childhood Treatment (PDQR): Health Professional Version. PDQ Pediatric Treatment Editorial Board. In: *PDQ Cancer Information Summaries (Internet)*. Bethesda (MD): National Cancer Institute (US); 2002-2017 Mar 2.

[50] Kurihara, N., Saito, H., Nanjo, H., Konno, H., Atari, M., Saito, Y., Fujishima, S., Kameyama, K. and Minamiya, Y. (2016). Thymic carcinoma with myas-thenia gravis: Two case reports. *Int J Surg Case Rep*, 27: 110-112.

[51] Huete-Garin, A., Sagel, S. S. (2005). Chapter 6: "Mediastinum, Thymic Neoplasm". In: J. K. T. Lee; Sagel, S.S., Stanlej, Heiken, J. P. *Computed Body Tomography with MRI Correlation*, Philadelphia: Lippincott Williams & Wilkins. 311-324.

[52] Jaffe, E. S., Harris, N. L. Vardiman, J. W., Campo, E., Arber, D. A. (2011). *Hematopathology* (1ˢᵗ ed.). Elsevier Saunders.

[53] Yeh, J. H., Lin, C. C., Chen, Y. K., Sung, F. C., Chiu, H. C. and Kao, C. H. (2014). Excessive risk of cancer and in particular lymphoid

malignancy in myasthenia gravis patients: a population-based cohort study. *Neuromuscul Disord*, 24(3): 245-9.

[54] Uner, A. H., Abali, H., Engin, H., Akyol, A., Ruacan, S., Tan, E., Güllü, I., Altundağ K. and Güler, N. (2001). Myasthenia gravis and lymphoblastic lymphoma involving the thymus: a rare association. *Leuk Lymphoma*, 42(3): 527-531.

[55] Chang, H., Chen, T. J., Chuang, W. Y. and Lin, T. L. (2011). Precursor B-cell acute lymphoblastic leukemia after thymoma and myasthenia gravis: report of a case and review of the literature. *Tumori*, 97(1): 126-129.

[56] Marx, A., Willcox, N., Leite, M. I., Chuang, W. Y., Schalke, B., Nix, W. and Strobel, P. (2010). Thymoma and paraneoplastic myasthenia gravis. *Autoimmunity*, 43: 413–427.

[57] Willcox, N., Leite, M. I., Kadota, Y., Jones, M., Meager, A., Subrahmanyam, P., Dasgupta, B., Morgan, B. P. and Vincent, A. (2008). Autoimmunizing mechanisms in thymoma and thymus. *Ann NY Acad Sci*, 1132: 163–173.

[58] Hapnes, L., Willcox, N., Oftedal, B. E., Owe, J. F., Gilhus, N. E., Meager, A., Husebye, E. S. and Wolff, A. S. (2012). Radioligand-binding assay reveals distinct autoantibody preferences for type I interferons in APS I and myasthenia gravis subgroups. *J Clin Immunol*, 32: 230–237.

[59] Kisand, K., Lilic, D., Casanova, J. L., Peterson, P., Meager, A. and Willcox, N. (2011). Mucocutaneous candidiasis and auto-immunity against cytokines in APECED and thymoma patients: clinical and pathogenetic implications. *Eur J Immunol*, 41:1517–1527.

[60] Tackenberg, B., Schlegel, K., Happel, M., Eienbroker, C., Gellert, K., Oertel, W. H., Meager, A., Willcox, N. and Sommer, N. (2009) Expanded TCR Vbeta subsets of CD8(+) T-cells in late-onset myasthenia gravis: novel parallels with thymoma patients. *J Neuroimmunol*, 216: 85–91.

[61] Binks, S., Vincent, A. and Palace, J. (2016). Myasthenia gravis: a clinical-immunological update. *J Neurol*, 263(4): 826-34.

[62] Hurst, R. L. and Gooch, C. L. (2016). Muscle-Specific Receptor Tyrosine Kinase (MuSK) Myasthenia Gravis. *Curr Neurol Neurosci Rep*, 16(7): 61.

[63] Suzuki, S., Utsugisawa, K., Nagane, Y. and Suzuki, N. (2011). Three types of striational antibodies in myasthenia gravis. *Autoimmune Dis*, 2011: 740583.

[64] Suzuki, S., Baba, A., Kaida, K., Utsugisawa, K., Kita, Y., Tsugawa, J., Ogawa, G., Nagane, Y., Kuwana, M. and Suzuki, N. (2014). Cardiac involvements in myasthenia gravis associated with anti Kv1.4 antibodies. *Eur J Neurol*, 21(2): 223-230.

[65] Cortés-Vicente E., Gallardo E., Martínez MÁ, Díaz-Manera J., Querol L., Rojas-García R. and Illa I. (2016). Clinical Characteristics of Patients with Double-Seronegative Myasthenia Gravis and Antibodies to Cortactin. *JAMA Neurol*, 73(9): 1099-1104.

[66] Yi, J. S., Russo, M. A., Massey, J. M., Juel, V., Hobson-Webb, L. D., Gable, K., Raja, S. M., Balderson, K., Weinhold, K. J. and Guptill, J. T. (2017). B10 Cell Frequencies and Suppressive Capacity in Myasthenia Gravis Are Associated with Disease Severity. *Front Neurol*, 10; 8: 34.

[67] Alahgholi-Hajibehzad, M., Oflazer, P., Aysa, F., Durmuş, H., Gülşen-Parman, Y., Marx, A., Deymeer, F. and Saruhan-Direskeneli, G. (2015). Regulatory function of CD4+CD25++ T cells in patients with myasthenia gravis is associated with phenotypic changes and STAT5 signaling: 1.25 Dihydroxyvitamin D3 modulates the suppressor activity. *J Neuroimmunol*, 15; 281: 51-60.

[68] Kohler, S., Keil, T. O., Swierzy, M., Hoffmann, S., Schaffert, H., Ismail, M., Rückert, J. C., Alexander, T., Hiepe, F., Gross, C., Thiel, A. and Meisel, A. (2013). Disturbed B cell subpopulations and increased plasma cells in myasthenia gravis patients. *J Neuroimmunol*, 15; 264(1-2): 114-9.

[69] Fan, X., Lin, C., Han, J., Jiang, X., Zhu, J. and Jin, T. (2015). Follicular Helper CD4+ T Cells in Human Neuroautoimmune Diseases and Their Animal Models. *Mediators Inflamm*, 2015: 638968.

[70]   Roche, J. C., Capablo, J. L., Larrad. L., Gervas-Arruga, J., Ara, J. R., Sánchez, A. and Alarcia, R. (2011). Increased serum interleukin-17 levels in patients with myasthenia gravis. *Muscle Nerve*, 44: 278–80.

[71]   Wang, Z., Wang, W., Chen, Y. and Wei, D. (2012). T helper type 17 cells expand in patients with myasthenia-associated thymoma. *Scand J Immunol*, 76(1): 54-61.

[72]   Yin, W., Allman, W., Ouyang, S., Li, Y., Li, J., Christadoss, P. and Yang, H. (2013). The increased expression of CD21 on AchR specified B cells in patients with myasthenia gravis. *J Neuroimmunol*, 15; 256(1-2): 49-54.

[73]   Lu, J., Li, J., Zhu, T. Q., Zhang, L., Wang, Y., Tian, F. F., Yang, H. (2013). Modulation of B cell regulatory molecules CD22 and CD72 in myasthenia gravis and multiple sclerosis. *Inflammation*, 36(3): 521-528.

[74]   Uzawa, A., Kanai, T., Kawaguchi, N., Oda, F., Himuro, K. and Kuwabara, S. (2016). Changes in inflammatory cytokine networks in myasthenia gravis. *Sci Rep*, 13; 6: 25886.

[75]   Molin, C. J., Westerberg, E. and Punga, A. R. (2017). Profile of upregulated inflammatory proteins in sera of Myasthenia Gravis patients. *Sci Rep*, 3; 7: 39716.

[76]   Yilmaz, V., Oflazer, P., Aysal, F., Durmus, H., Poulas, K., Yentur, S. P., Gulsen-Parman, Y., Tzartos, S., Marx, A., Tuzun, E., Deymeer, F. and Saruhan-Direskeneli, G. (2015). Differential Cytokine Changes in Patients with Myasthenia Gravis with Antibodies against AChR and MuSK. *PLoS One*, 20; 10(4): e0123546.

[77]   Zhang, D. Q., Wang, R., Li, T., Li, X., Qi, Y., Wang, J. and Yang, L. (2015). Remarkably increased resistin levels in anti-AChR antibody-positive myasthenia gravis. *J Neuroimmunol*, 283: 7–10.

[78]   Yang, D., Su, Z., Wu, S., Bi, Y., Li, X., Li, J., Lou, K., Zhang, H. and Zhang, X. (2016). Low antioxidant status of serum bilirubin, uric acid, albumin and creatinine in patients with myasthenia gravis. *Int J Neurosci*, 126(12): 1120-1126.

[79]   Punga, A. R., Andersson, M., Alimohammadi, M. and Punga, T. (2015). Disease specific signature of circulating miR-150-5p and

miR-21-5p in myasthenia gravis patients. *J Neurol Sci*, 15; 356(1-2): 90-96.

[80]  Punga, T., Bartoccioni, E., Lewandowska, M., Damato, V., Evoli, A. and Punga, A. R. (2016). Disease specific enrichment of circulating let-7 family microRNA in MuSK+ myasthenia gravis. *J Neuroimmunol*, 15; 292: 21-26.

[81]  Bartoccioni, E., Scuderi, F., Augugliaro, A., Chiatamone Ranieri, S., Sauchelli, D., Alboino, P., Marino, M., Evoli, A. (2009). HLA class II allele analysis in MuSK-positive myasthenia gravis suggests a role for DQ5. *Neurology*, 13; 72(2): 195-197.

[82]  Niks EH, Kuks JB, Roep BO, Haasnoot GW, Verduijn W., Ballieux B. E., De Baets M. H., Vincent A. and Verschuuren J. J. (2006). Strong association of MuSK antibody-positive myasthenia gravis and HLA-DR14-DQ5. *Neurology*, 13; 66(11): 1772-4.

[83]  Gregersen, P. K., Kosoy, R., Lee, A. T., Lamb, J., Sussman, J., McKee, D., Simpfendorfer, K. R., Pirskanen-Matell, R., Piehl, F., Pan-Hammarstrom, Q., Verschuuren, J. J., Titulaer, M. J., Niks, E. H., Marx, A., Ströbel, P., Tackenberg, B., Pütz, M., Maniaol, A., Elsais, A., Tallaksen, C., Harbo, H. F., Lie, B. A., Raychaudhuri, S., de Bakker, P. I., Melms, A., Garchon, H. J., Willcox, N., Hammarstrom, L. and Seldin, M. F. (2012). Risk for myasthenia gravis maps to a (151) Pro→Ala change in TNIP1 and to human leukocyte antigen-B*08. *Ann Neurol*, 72(6): 927-935.

[84]  Renton, A. E., Pliner, H. A., Provenzano, C., Evoli, A., Ricciardi, R., Nalls, M. A., Marangi, G., Abramzon, Y., Arepalli, S., Chong, S., Hernandez, D. G., Johnson, J. O., Bartoccioni, E., Scuderi, F., Maestri, M., Gibbs, J. R., Errichiello, E., Chiò, A., Restagno, G., Sabatelli, M., Macek, M., Scholz, S. W., Corse, A., Chaudhry, V., Benatar, M., Barohn, R. J., McVey, A., Pasnoor, M., Dimachkie, M. M., Rowin, J., Kissel, J., Freimer, M., Kaminski, H. J., Sanders, D. B., Lipscomb, B., Massey, J. M., Chopra, M., Howard, J. F. Jr, Koopman, W. J., Nicolle, M. W., Pascuzzi, R. M. Pestronk, A., Wulf, C., Florence, J., Blackmore, D., Soloway, A., Siddiqi, Z., Muppidi, S., Wolfe, G., Richman, D., Mezei, M. M., Jiwa, T., Oger, J.,

Drachman, D. B. and Traynor, B. J. (2015). A genome-wide association study of myasthenia gravis. *JAMA Neurol*, 72(4): 396-404.

[85] Papke, R. L., Wecker L. and Stitzel, J. A. (2010). Activation and inhibition of mouse muscle and neuronal nicotinic acetylcholine receptors expressed in Xenopus oocytes. *J Pharmacol Exp Ther*, 333(2): 501-18.

[86] Unwin, N. (2013). Nicotinic acetylcholine receptor and the structural basis of neuromuscular transmission: insights from Torpedo postsynaptic membranes. *Q Rev Biophys*, 46(4): 283-322.

[87] Miyazawa, A., Fujiyoshi, Y. and Unwin, N. (2003). Structure and gating mechanism of the acetylcholine receptor pore. *Nature*, 423:949–955.

[88] Albuquerque, E. X., Pereira, E. F., Alkondon, M. and Rogers, S. W. (2009). Mammalian nicotinic acetylcholine receptors: from structure to function. *Physiol Rev*, 89: 73–120.

[89] Chakrapani, S. and Auerbach, A. (2005). Remarkably increased resistin levels in anti-AChR antibody-positive myasthenia gravis *Proc of the Natl Acad of Sci USA*, 4; 102: 87–92.

[90] Drachman, D. B., de Silva, S., Ramsay, D. and Pestronk, A. (1987). Humoral pathogenesis of myasthenia gravis. *Ann N Y Acad Sci*, 505: 90–105.

[91] Engel, A. G. (1979). The immunopathological basis of acetylcholine receptor deficiency in myasthenia gravis. *Prog Brain Res*, 49: 423–434.

[92] Burges, J., Wray, D. W., Pizzighella, S., Hall, Z., Vincent, A. (1990). A myasthenia gravis plasma immunoglobulin reduces miniature endplate potentials at human endplates *in vitro*. *Muscle Nerve*, 13: 407–413.

[93] Howard, F. M. Jr., Lennon, V. A., Finley, J., Matsumoto, J. and Elveback, L. R. (1987). Clinical correlations of antibodies that bind, block, or modulate human acetylcholine receptors in myasthenia gravis. *Ann NY Acad Sci*, 505: 526–538.

[94] Tüzün, E., Scott, B. G., Goluszko, E., Higgs, S. and Christadoss, P. (2003). Genetic evidence for involvement of classical complement pathway in induction of experimental autoimmune myasthenia gravis. *J Immunol*, 1;171(7): 3847-3854.

[95] Christadoss, P., Tüzün, E., Li, J., Saini, S. S. and Yang, H. (2008). Classical complement pathway in experimental autoimmune myasthenia gravis pathogenesis. *Ann N Y Acad Sci*, 1132: 210–219.

[96] Luo, J. and Lindstrom, J. (2012). Myasthenogenicity of the main immunogenic region and endogenous muscle nicotinic acetyl-choline receptors. *Autoimmunity*, 45: 245–252.

[97] Kordas, G., Lagoumintzis, G., Sideris, S., Poulas, K. and Tzartos, S. J. (2014) Direct Proof of the *In vivo* Pathogenic Role of the AChR Autoantibodies from Myasthenia Gravis Patients. *PLoS One*, 9(9): e108327.

[98] Tzartos, S. J., Seybold, M. E. and Lindstrom, J. M. (1982). Specificities of antibodies to acetylcholine receptors in sera from myasthenia gravis patients measured by monoclonal antibodies. *Proc Natl Acad Sci U S A*, 79: 188–192.

[99] Ragheb, S., Mohamed, M. and Lisak, R. P. (2005). Myasthenia gravis patients, but not healthy subjects, recognize epitopes that are unique to the epsilon-subunit of the acetylcholine receptor. *J Neuroimmunol*, 159: 137–145.

[100] Masuda, T., Motomura, M., Utsugisawa, K., Nagane, Y., Nakata, R., Tokuda, M., Fukuda, T., Yoshimura, T., Tsujihata, M. and Kawakami, A. (2012). Antibodies against the main immunogenic region of the acetylcholine receptor correlate with disease severity in myasthenia gravis. *J Neurol Neurosurg Psychiatry*, 83(9): 935-940.

[101] Vincent, A., Jacobson, L., Shillito, P. (1994). Response to human acetylcholine receptor alpha 138–199: determinant spreading initiates autoimmunity to self-antigen in rabbits. *Immunol Lett*, 39: 269–275.

[102] Curnow, J., Corlett, L., Willcox, N. and Vincent, A. (2001). Presentation by myoblasts of an epitope from endogenous

acetylcholine receptor indicates a potential role in the spreading of the immune response. *J Neuroimmunol*, 115: 127–134.

[103] Maclennan, C. A., Vincent, A., Marx, A., Willcox, N., Gilhus, N. E., Newsom-Davis J. and Beeson, D. (2008). Preferential expression of AChR epsilon-subunit in thymomas from patients with myasthenia gravis. *J Neuroimmunol*, 201–202: 28–32.

[104] Gilhus, N. E., Skeie, G. O., Romi, F., Lazaridis, K., Zisimopoulou, P., Tzartos, S. (2016). Myasthenia gravis - autoantibody characteristics and their implications for therapy. *Nat Rev Neurol*, 12(5): 259-68.

[105] Patrick, J. and Lindstrom, J. (1973). Autoimmune response to acetylcholine receptor. *Science*, 180 (4088): 871–872.

[106] Toyka, K. V., Brachman, D. B., Pestronk, A., Kao, I. (1975). Myasthenia gravis: passive transfer from man to mouse. *Science*, 190 (4212): 397–399.

[107] Phillips, W. D. and Vincent, A. (2016). Pathogenesis of myasthenia gravis: update on disease types, models, and mechanisms. *F1000Res*, 27; 5. pii: F1000 Faculty Rev-1513.

[108] Vincent, A. and Newsom-Davis, J. (1985). Acetylcholine receptor antibody as a diagnostic test for myasthenia gravis: results in 153 validated cases and 2967 diagnostic assays. *J. Neurol. Neurosurg. Psychiatry*, 48: 1246–1252.

[109] Oger, J., Kaufman, R. and Berry, K. (1987). Acetylcholine receptor antibodies in myasthenia gravis: use of a qualitative assay for diagnostic purposes. *Can J Neurol Sci*, 14 (3): 297–302.

[110] Oger, J. J. (1993). Thymus histology and acetylcholine receptor antibodies in generalized myasthenia gravis. *Ann N Y Acad Sci*, 21; 681: 110.

[111] Rigamonti, A., Lauria, G., Piamarta, F., Fiumani, A. and Agostoni E. (2011). Thymoma-associated myasthenia gravis without acetyl-choline receptor antibodies. *J Neurol Sci,* 15; 302(1-2): 112-113.

[112] Meriggioli, M. N. and Sanders, D. B. (2012). Muscle autoantibodies in myasthenia gravis: beyond diagnosis? *Expert Rev Clin Immunol*, 8(5): 427-438.

[113] Patrick, J., Lindstrom, J., Culp, B. and McMillan, J. (1973). Studies on purified eel acetylcholine receptor and anti-acetylcholine receptor antibody. *Proc Natl Acad Sci U S A*, 70(12): 3334–3338.

[114] Shi, Q. G., Wang, Z. H., Ma, X. W., Zhang, D. Q., Yang, C. S., Shi, F. D. and Yang, L. (2012). Clinical significance of detection of antibodies to fetal and adult acetylcholine receptors in myasthenia gravis. *Neurosci Bull*, 28(5): 469-74.

[115] Kawanami, S., Tsuji, R. and Oda, K. (1984). Enzyme-linked immunosorbent assay for antibody against the nicotinic acetylcholine receptor in human myasthenia gravis. *Ann. Neurol*, 15: 195–200.

[116] Kaufman, R. L., Oger, J. (1988). Antibody production by blood lymphocytes in myasthenia gravis: reduction in disease of long duration. *Neurology*, 38(5): 818–821.

[117] Masuda, T., Motomura, M., Utsugisawa, K., Nagane, Y., Nakata, R., Tokuda, M., Fukuda, T., Yoshimura, T., Tsujihata, M. and Kawakami, A. (2012). Antibodies against the main immunogenic region of the acetylcholine receptor correlate with disease severity in myasthenia gravis. *J Neurol Neurosurg Psychiatry*, 83(9): 935-940.

[118] Yeh, J. H., Chen, W. H., Chiu, H. C. (2003). Predicting the course of myasthenic weakness following double filtration plasmapheresis. *Acta Neurol Scand*, 108(3): 174-178.

[119] Sanders, D. B., Burns, T. M., Cutter, G. R., Massey, J. M., Juel, V. C., Hobson-Webb, L; Muscle Study Group. (2014). Does change in acetylcholine receptor antibody level correlate with clinical change in myasthenia gravis? *Muscle Nerve,* 49(4): 483-486.

[120] Gilhus, N. E., Skeie, G. O., Romi, F., Lazaridis, K. Zisimopoulos, P. Tzartos, S. (2016). Myasthenia gravis – autoantibody characteristics and their implications for therapy. *Nat Rev Neurol*, 12(5): 259-268.

[121] Leite, M. I., Jacob, S., Viegas, S., Cossins, J., Clover, L., Morgan, B. P., Beeson, D., Willcox, N. and Vincent, A. (2008). IgG1 antibodies to acetylcholine receptors in "seronegative" myasthenia gravis. *Brain*, 131(Pt7): 1940–1952.

[122] Vincent, A., Waters, P., Leite, M. I., Jacobson, L., Koneczny, I., Cossins, J. and Beeson D. (2012). Antibodies identified by cell-based

assays in myasthenia gravis and associated diseases. *Ann. N. Y. Acad. Sci*, 1274: 92–98.

[123] Devic, P., Petiot, P., Simonet, T., Stojkovic, T., Delmont, E., Franques, J., Magot, A., Vial, C., Lagrange, E., Nicot, A. S., Risson, V., Eymard B. and Schaeffer, L. (2014). Antibodies to clustered acetylch, oline receptor: expanding the phenotype. *Eur J Neurol*, 21(1): 130-134.

[124] Jacob, S., Viegas, S., Leite, M. I., Webster, R., Cossins, J., Kennett, R., Hilton-Jones, D., Morgan, B. P. and Vincent. A. (2012). Presence and pathogenic relevance of antibodies to clustered acetylcholine receptor in ocular and generalized myasthenia gravis. *Arch. Neurol*, 69: 994–1001.

[125] Rodriguez Cruz, P. M., Huda, S., López-Ruiz, P. and Vincent, A. (2015). Use of cell-based assays in myasthenia gravis and other antibody-mediated diseases. *Exp Neurol*, 270: 66-71.

[126] Klein, R., Marx, A., Ströbel, P., Schalke, B., Nix, W., Willcox, N. (2013). Autoimmune associations and autoantibody screening show focused recognition in patient subgroups with generalized myasthenia gravis. *Hum Immunol*, 74(9): 1184-1193.

[127] Waters, P. and Vincent, A. (2008). Detection of anti-aquaporin-4 antibodies in neuromyelitis optica: Current status of the assays. *Int MS J/MS Forum*, 15(3): 99–105.

[128] Freitas, E. and Guimarães, J. (2015). Neuromyelitis optica spectrum disorders associated with other autoimmune diseases. *Rheumatol Int*, 35(2): 243-53.

[129] Otsuka, K., Ito, M., Ohkawara, B., Masuda, A., Kawakami, Y., Sahashi, K., Nishida, H., Mabuchi, N., Takano, A., Engel, A. G. and Ohno, K. (2015). Collagen Q and anti-MuSK autoantibody competitively suppress agrin/LRP4/MuSK signaling. *Sci Rep*, 10; 5: 13928.

[130] Okada, K., Inoue, A., Okada, M., Murata, Y., Kakuta, S., Jigami, T., Kubo, S., Shiraishi, H., Eguchi, K., Motomura, M., Akiyama, T., Iwakura, Y., Higuchi, O. and Yamanashi, Y. (2006). The muscle

protein Dok-7 is essential for neuromuscular synaptogenesis. *Science*, 312: 1802–1805.

[131] DeChiara, T. M., Bowen, D. C., Valenzuela, D. M., Simmons, M. V., Poueymirou, W. T., Thomas, S., Kinetz, E., Compton, D. L., Rojas, E., Park, J. S., Smith, C., DiStefano, P. S., Glass, D. J., Burden, S. J. and Yancopoulos, G. D. (1996). The receptor tyrosine kinase MuSK is required for neuromuscular junction formation *in vivo*. *Cell*, 85(4): 501–512.

[132] Kim, N., Stiegler, A. L., Cameron, T. O., Hallock, P. T., Gomez, A. M., Huang, J. H., Hubbard, S. R., Dustin, M. L. and Burden, S. J. (2008). Lrp4 is a receptor for Agrin and forms a complex with MuSK. *Cell*, 135(2): 334–342.

[133] Zhang, B., Luo, S., Wang, Q., Suzuki, T., Xiong, W. C., Mei, L. LRP4 serves as a coreceptor of agrin. *Neuron*. 2008; 60(2): 285–297.

[134] Ghazanfari, N., Fernandez, K. J., Murata, Y., Morsch, M., Ngo, S. T., Reddel S. W., Noakes, P. G., Phillips, W. D. (2011). Muscle specific kinase: organizer of synaptic membrane domains. *Int J Biochem Cell Biol*, 43(3): 295–298.

[135] Cheusova, T., Khan, M A., Schubert, S. W., Gavin, A. C., Buchou, T., Jacob, G., Sticht, H., Allende, J., Boldyreff, B., Brenner, H. R. and Hashemolhosseini, S. (2006). Casein kinase 2-dependent serine phosphorylation of MuSK regulates acetylcholine receptor aggregation at the neuromuscular junction. *Genes Dev*, 20(13):1800-1816.

[136] Hoch, W., McConville, J., Helms, S., Newsom-Davis, J., Melms, A. and Vincent, A. (2001). Autoantibodies to the receptor tyrosine kinase MuSK in patients with myasthenia gravis without acetylcholine receptor antibodies. *Nat Med*, 7: 365-368.

[137] Huijbers, M. G., Zhang, W., Klooster, R., Niks, E. H., Friese, M. B., Straasheijm, K. R., Thijssen, P. E., Vrolijk, H., Plomp, J. J., Vogels, P., Losen, M., Van der Maarel, S. M., Burden, S. J. and Verschuuren, J. J. (2013). MuSK IgG4 autoantibodies cause myasthenia gravis by inhibiting binding between MuSK and Lrp4. *Proc. Natl. Acad. Sci. U. S. A*, 110: 20783–20788.

[138] Koneczny, I., Cossins, J., Waters, P., Beeson, D. and Vincent, A. (2013). MuSK myasthenia gravis IgG4 disrupts the interaction of LRP4 with MuSK but both IgG4 and IgG1-3 can disperse preformed agrin-independent AChR clusters. *PLoS One*, 8: e80695.

[139] Ghazanfari, N., Morsch M., Reddel, S. W., Liang, S. X. and Phillips, W. D. (2014). Muscle-specific kinase (MuSK) autoantibodies suppress the MuSK pathway and ACh receptor retention at the mouse neuromuscular junction. *J Physiol*, 592(13): 2881–2897.

[140] Ghazanfari, N., Linsao, E. L., Trajanovska, S., Morsch, M., Gregorevic, P., Liang, S. X., Reddel, S. W. and Phillips, W. D. (2015). Forced expression of muscle specific kinase slows postsynaptic acetylcholine receptor loss in a mouse model of MuSK myasthenia gravis. *Physiol Rep,* 3(12): pii: e12658.

[141] Morsch, M., Reddel, S. W., Ghazanfari, N., Toyka, K. V. and Phillips, W. D. (2012). Muscle specific kinase autoantibodies cause synaptic failure through progressive wastage of postsynaptic acetylcholine receptors. *Exp Neurol,* 237(2): 286–295.

[142] Kawakami, Y., Ito, M., Hirayama, M., Sahashi, K., Ohkawara, B., Masuda, A., Nishida, H., Mabuchi, N., Engel, A. G. and Ohno, K. (2011). Anti-MuSK autoantibodies block binding of collagen Q to MuSK. *Neurology*, 77, 1819–1826.

[143] Shigemoto, K., Kubo, S., Maruyama, N., Hato, N., Yamada, H., Jie, C., Kobayashi, N., Mominoki, K., Abe, Y., Ueda, N. and Matsuda, S. (2006). Induction of myasthenia by immunization against muscle-specific kinase. *J Clin Invest*, 116(4): 1016–1024.

[144] Jha, S., Xu, K., Maruta, T., Oshima, M., Mosier, D. R., Atassi, M. Z. and Hoch, W. (2006). Myasthenia gravis induced in mice by immunization with the recombinant extracellular domain of rat muscle-specific kinase (MuSK). *J Neuroimmunol*, 175(1–2): 107–117.

[145] Punga, A. R., Lin, S., Oliveri, F., Meinen, S., Rüegg, M. A. (2011). Muscle-selective synaptic disassembly and reorganization in MuSK antibody positive MG mice. *Exp Neurol,* 230(2): 207–217.

[146] Richman, D. P., Nishi, K., Morell, S. W., Chang, J. M., Ferns, M. J., Wollmann, R. L., Maselli, R. A., Schnier, J. and Agius, M. A. (2012). Acute severe animal model of antimuscle-specific kinase myasthenia: combined postsynaptic and presynaptic changes. *Arch Neurol*, 69(4): 453–460.

[147] Mori, S., Kubo, S., Akiyoshi, T., Yamada, S., Miyazaki, T., Hotta, H., Desaki, J., Kishi, M., Konishi, T., Nishino, Y., Miyazawa, A., Maruyama, N. and Shigemoto, K. (2012). Antibodies against muscle-specific kinase impair both presynaptic and postsynaptic functions in a murine model of myasthenia gravis. *Am J Pathol*, 180(2): 798-810.

[148] Patel, V., Oh, O., Voit, A., Sultatos, L. G., Babu, G. J., Wilson BA3, Ho M3, McArdle JJ1. (2014). Altered active zones, vesicle pools, nerve terminal conductivity, and morphology during experimental MuSK myasthenia gravis. *PLoS One*, 9(12): e110571.

[149] Cole, R. N., Reddel, S. W., Gervásio, O. L. and Phillips, W. D. (2008). Anti-MuSK patient antibodies disrupt the mouse neuromuscular junction. *Ann Neurol*, 63(6): 782–789.

[150] Cole, R. N., Ghazanfari, N., Ngo, S. T., Gervásio, O. L., Reddel, S. W. and Phillips, W. D. (2010). Patient autoantibodies deplete postsynaptic muscle-specific kinase leading to disassembly of the Ach receptor scaffold and myasthenia gravis in mice. *J Physiol.*, 588(Pt 17): 3217–3229.

[151] Klooster, R., Plomp, J. J., Huijbers, M. G., Niks, E. H., Straasheijm, K. R., Detmers, F. J., Hermans, P. W., Sleijpen, K., Verrips, A., Losen, M., Martinez-Martinez, P., De Baets, M. H., van der Maarel, S. M. and Verschuuren, J. J. (2012). Muscle-specific kinase myasthenia gravis IgG4 autoantibodies cause severe neuromuscular junction dysfunction in mice. *Brain,* 135(Pt 4): 1081–1101.

[152] Viegas, S., Jacobson, L., Waters, P., Cossins, J., Jacob, S., Leite, M. I., Webster, R. and Vincent, A. (2012). Passive and active immunization models of MuSK-Ab positive myasthenia: Electro-physiological evidence for pre and postsynaptic defects. *Exp Neurol.*, 234(2): 506–512.

[153] Selcen, D., Fukuda, T., Shen, X. M. and Engel, A. G. (2004). Are MuSK antibodies the primary cause of myasthenic symptoms? *Neurology,* 62: 1945–1950.

[154] Shiraishi H., Motomura M., Yoshimura T., Fukudome T., Fukuda T., Nakao Y, Tsujihata M., Vincent, A. and Eguchi, K. (2005). Acetylcholine receptors loss and postsynaptic damage in MuSK antibody-positive myasthenia gravis. *Ann Neurol*, 57: 289–293.

[155] Punga, A. R., Maj, M., Lin, S., Meinen, S. and Ruegg, M. A. (2011). MuSK levels differ between adult skeletal muscles and influence postsynaptic plasticity. *Eur J Neurosci,* 33: 890–898.

[156] McConville, J., Farrugia, M. E., Beeson, D., Kishore, U., Metcalfe, R., Newsom-Davis, J. and Vincent, A. (2004). Detection and characterization of MuSK antibodies in seronegative myasthenia gravis. *Ann Neurol*, 55(4): 580–584.

[157] Nirula, A., Glaser, S. M., Kalled, S. L. and Taylor, F. R. (2011). What is IgG4? A review of the biology of a unique immunoglobulin subtype. *Curr Opin Rheumatol*, 23(1): 119-124.

[158] van der Neut Kolfschoten, M., Schuurman, J., Losen, M., Bleeker, W. K., Martínez-Martínez, P., Vermeulen, E., den Bleker, T. H., Wiegman, L., Vink, T., Aarden, L. A., De Baets, M. H., van de Winkel, J. G., Aalberse, R. C. and Parren, P. W. (2007). Anti-inflammatory activity of human IgG4 antibodies by dynamic Fab arm exchange. *Science*, 14;317(5844):1554-7.

[159] Aalberse, R. C., Stapel, S. O., Schuurman, J. and Rispens, T. (2009). Immunoglobulin G4: an odd antibody. *Clin Exp Allergy*, 39(4):469-477.

[160] Klooster, R., Plomp, J. J., Huijbers, M. G., Niks, E. H., Straasheijm, K. R., Detmers, F. J., Hermans, P. W., Sleijpen, K., Verrips, A., Losen, M., Martinez-Martinez, P., De Baets, M. H., van der Maarel, S. M. and Verschuuren, J. J. (2001). Muscle-specific kinase myasthenia gravis IgG4 autoantibodies cause severe neuromuscular junction dysfunction in mice. *Brain,* 135(Pt 4):1081-1101.

[161] Koneczny, I., Cossins, J., Waters, P., Beeson, D. and Vincent, A. (2013). MuSK myasthenia gravis IgG4 disrupts the interaction of

LRP4 with MuSK but both IgG4 and IgG1-3 can disperse preformed agrin-independent AChR clusters. *PLoS One*, 7; 8(11): e80695.

[162] Küçükerden, M., Huda, R., Tüzün, E., Yılmaz, A., Skriapa, L., Trakas, N., Strait, R. T., Finkelman, F. D., Kabadayı, S., Zisimopoulou, P., Tzartos, S. and Christadoss, P. (2016). MuSK induced experimental autoimmune myasthenia gravis does not require IgG1 antibody to MuSK. *J Neuroimmunol*, 15;295-296:84-92.

[163] Benveniste, O., Jacobson, L., Farrugia, M. E., Clover, L. and Vincent, A. (2005). MuSK antibody positive myasthenia gravis plasma modifies MURF-1 expression in C2C12 cultures and mouse muscle *in vivo*. *J Neuroimmunol*, 170:41–48.

[164] Boneva, N., Frenkian-Cuvelier, M., Bidault, J., Brenner, T. and Berrih-Aknin, S. (2006). Major pathogenic effects of anti-MuSK antibodies in myasthenia gravis. *J Neuroimmunol* 177:119–131.

[165] Farrugia, M. E., Bonifati, D. M., Clover, L., Cossins, J., Beeson, D. and Vincent, A. (2007). Effect of sera from AChR-antibody negative myasthenia gravis patients on AChR and MuSK in cell cultures. *J Neuroimmunol*, 185:136–144.

[166] Evoli, A., Tonali, P. A., Padua, L., Monaco, M. L., Scuderi, F., Batocchi, A. P., Marino, M. and Bartoccioni, E. (2003). Clinical correlates with anti-MuSK antibodies in generalized seronegative myasthenia gravis. *Brain*, 126(Pt 10):2304–2311.

[167] Gilhus, N. E. (2012). Myasthenia and the neuromuscular junction. *Curr Opin Neurol*, 25(5):523-529.

[168] Kostera-Pruszczyk, A., Kamińska, A., Dutkiewicz, M., Emeryk-Szajewska, B., Strugalska-Cynowska, M. H., Vincent, A. and Kwieciński, H. (2008). MuSK-positive myasthenia gravis is rare in the Polish population. *Eur. J. Neurol*, 15(7):720-724.

[169] Vincent, A. (2008). Autoantibodies in neuromuscular transmission disorders. *Ann. Indian Acad. Neurol*, 11: 140–145.

[170] Yeh, J. H., Chen, W. H., Chiu, H. C. and Vincent, A. (2004). Low frequency of MuSK antibody in generalized seronegative myasthenia gravis among Chinese. *Neurology*, 62(11):2131–2132.

[171] Tsonis, A. I., Zisimopoulou, P., Lazaridis, K., Tzartos, J., Matsigkou, E., Zouvelou, V., Mantegazza, R., Antozzi, C., Andreetta, F., Evoli, A., Deymeer, F., Saruhan-Direskeneli, G., Durmus, H., Brenner, T., Vaknin, A., Berrih-Aknin, S., Behin, A., Sharshar, T., De Baets, M., Losen, M., Martinez-Martinez, P., Kleopa, K. A., Zamba-Papanicolaou, E., Kyriakides, T., Kostera-Pruszczyk. A., Szczudlik, P., Szyluk, B., Lavrnic, D., Basta, I., Peric, S., Tallaksen, C., Maniaol, A., Casasnovas Pons, C., Pitha, J., Jakubíkova, M., Hanisch, F. and Tzartos, S. J. (2015). MuSK autoantibodies in myasthenia gravis detected by cell based assay—A multinational study. *J Neuroimmunol*, 15;284:10-17.

[172] Guptill, J. T., Sanders, D. B. and Evoli, A. (2011). Anti-MuSK antibody myasthenia gravis: clinical findings and response to treatment in two large cohorts. *Muscle Nerve*, 44:36-40.

[173] Ito, A., Sasaki, R., Ii, Y., Nakayama, S., Motomura, M. and Tomimoto, H. (2013). A case of thymoma-associated myasthenia gravis with anti-MuSK antibodies. *Rinsho Shinkeigaku*, 53(5):372-375.

[174] Ohta, K., Shigemoto, K., Kubo, S., Maruyama, N., Abe, Y., Ueda, N., Fujinami, A. and Ohta. (2005). MuSK Ab described in seropositive MG sera found to be Ab to alkaline phosphatase. *Neurology*, 27;65(12):1988.

[175] Bartoccioni, E., Scuderi, F., Minicuci, G. M., Marino, M., Ciaraffa, F. and Evoli, A. (2006). Anti-MuSK antibodies: correlation with myasthenia gravis severity. *Neurology*, 67(3):505.

[176] Guptill, J. T. and Sanders, D. B. (2010). Update on muscle-specific tyrosine kinase antibody positive myasthenia gravis. *Curr Opin Neurol*, 23:530–535.

[177] Kaminski, H. J., Kusner, L. L., Wolfe, G. I., Aban, I., Minisman, G., Conwit, R. and Cutter, G. (2012). Biomarker development for myasthenia gravis. *Ann N Y Acad Sci*, 1275:101-106.

[178] Niks, E. H., van Leeuwen, Y., Leite, M. I., Dekker, F. W., Wintzen, A. R., Wirtz, P. W., Vincent, A., van Tol, M. J., Jol-van der Zijde, C. M. and Verschuuren, J. J. (2008). Clinical fluctuations in MuSK

myasthenia gravis are related to antigen-specific IgG4 instead of IgG1. *Neuroimmunol,* 195(1-2):151-156.

[179] Shen, C., Lu, Y., Zhang, B., Figueiredo, D., Bean, J., Jung, J., Wu, H., Barik, A., Yin, D. M., Xiong, W. C. and Mei, L. (2013). Antibodies against low-density lipoprotein receptor-related protein 4 induce myasthenia gravis. *J Clin Invest,* 123(12):5190-5202.

[180] Simon-Chazottes, D., Tutois, S., Kuehn, M., Evans, M., Bourgade, F., Cook, S., Davisson, M. T. and Guenet, J. L. (2006). Mutations in the gene encoding the low-density lipoprotein receptor LRP4 cause abnormal limb development in the mouse. *Genomics,* 87:673-677.

[181] Choi, H. Y., Dieckmann, M., Herz, J. and Niemeier, A. (2009). Lrp4, a novel receptor for Dickkopf 1 and sclerostin, is expressed by osteoblasts and regulates bone growth and turnover *in vivo. PLoS One,* 4:e7930.

[182] Karner, C. M., Dietrich, M. F., Johnson, E. B., Kappesser, N., Tennert, C., Percin, F., Wollnik, B., Carroll, T. J. and Herz, J. (2010). Lrp4 regulates initiation of ureteric budding and is crucial for kidney formation–a mouse model for Cenani-Lenz syndrome. *PLoS One,* 5:e10418.

[183] Johnson, E. B., Hammer, R. E. and Herz, J. (2005). Abnormal development of the apical ectodermal ridge and polysyndactyly in Megf7-deficient mice. *Hum Mol Genet,* 14:3523–3538.

[184] Zhang, W., Coldefy, A. S., Hubbard, S. R. and Burden, S. J. (2011). Agrin binds to the N-terminal region of Lrp4 protein and stimulates association between Lrp4 and the first immunoglobulin-like domain in muscle-specific kinase (MuSK). *J Biol Chem,* 25;286(47):40624-40630.

[185] Yumoto, N., Kim, N. and Burden, S. J. (2012). Lrp4 is a retrograde signal for presynaptic differentiation at neuromuscular synapses. *Nature,* 20;489(7416):438-442.

[186] Shen, C., Lu, Y., Zhang, B., Figueiredo, D., Bean, J., Jung, J., Wu, H., Barik, A., Yin, D. M., Xiong, W. C., Mei, L. (2013). Antibodies against low-density lipoprotein receptor-related protein 4 induce myasthenia gravis. *J Clin Invest,* 123(12):5190-5202.

[187] Ulusoy, C., Çavuş, F., Yılmaz, V. and Tüzün, E. (2017). Immunization with Recombinantly Expressed LRP4 Induces Experimental Autoimmune Myasthenia Gravis in C57BL/6 Mice. *Immunol Invest*, 4:1-10.

[188] Gilhus, N. E. and Verschuuren, J. J. (2015). Myasthenia gravis: subgroup classification and therapeutic strategies. *Lancet Neurol*, 14(10):1023-36.

[189] Higuchi, O., Hamuro, J., Motomura, M. and Yamanashi, Y. (2011). Autoantibodies to low-density lipoprotein receptor-related protein 4 in myasthenia gravis. *Ann Neurol*, 6 9:418 – 422.

[190] Gomez, A. M. and Burden, S. J. (2011). The extracellular region of Lrp4 is sufficient to mediate neuromuscular synapse formation. *Dev Dyn*, 240(12):2626-2633.

[191] Pevzner, A., Schoser, B., Peters, K., Cosma, N. C., Karakatsani, A., Schalke, B., Melms, A. and Kröger, S. (2012). Anti-LRP4 autoantibodies in AChR- and MuSK-antibody-negative myasthenia gravis. *J Neurol*, 259(3):427-35.

[192] Zhang, B., Tzartos, J. S., Belimezi, M., Ragheb, S., Bealmear, B., Lewis, R. A., Xiong, W. C., Lisak, R. P., Tzartos, S. J. and Mei, L. (2012). Autoantibodies to lipoprotein-related protein 4 in patients with double-seronegative myasthenia gravis. *Arch Neurol*, 69(4): 445-51.

[193] Tzartos, J. S., Zisimopoulou, P., Rentzos, M., Karandreas, N., Zouvelou, V., Evangelakou, P., Tsonis, A., Thomaidis, T., Lauria, G., Andreetta, F., Mantegazza, R. and Tzartos, S. J. (2014). LRP4 antibodies in serum and CSF from amyotrophic lateral sclerosis patients. *Ann clin transl neurol*, 1(2):80–87.

[194] Tzartos, J., Zisimopoulou, P., Tsonis, A., Evangelakou, P., Rentzos, M. and Karandreas, N., (2015). LRP4 antibodies are frequent in serum and CSF from amyotrophic lateral sclerosis patients (S34. 004). *Neurology*, 84(suppl).

[195] Rivner, M. H., Liu, S., Quarles, B., Fleenor, B., Shen, C. and Pan, J. (2017). Agrin and low-density lipoprotein-related receptor protein 4

antibodies in amyotrophic lateral sclerosis patients. *MeiL Muscle Nerve*, 55(3):430-432.

[196] Takahashi, H., Noto, Y. I., Makita, N., Kushimura-Okada, Y., Ishii, R., Tanaka, A., Ohara, T., Nakane, S., Higuchi, O., Nakagawa, M. and Mizuno, T. (2016). Myasthenic symptoms in anti-low-density lipoprotein receptor-related protein 4 antibody-seropositive amyotrophic lateral sclerosis: two case reports. *BMC Neurol*, 18; 16(1):229.

[197] Zisimopoulou, P., Evangelakou, P., Tzartos, J., Lazaridis, K., Zouvelou, V., Mantegazza, R., Antozzi, C., Andreetta, F., Evoli, A., Deymeer, F., Saruhan-Direskeneli, G., Durmus, H., Brenner, T., Vaknin, A., Berrih-Aknin, S., Frenkian Cuvelier, M., Stojkovic, T., DeBaets, M., Losen, M., Martinez-Martinez, P., Kleopa, K. A. 1, Zamba-Papanicolaou, E., Kyriakides, T., Kostera-Pruszczyk, A., Szczudlik, P., Szyluk, B., Lavrnic, D., Basta, I., Peric, S., Tallaksen, C., Maniaol, A. and Tzartos, S. J. (2014). A comprehensive analysis of the epidemiology and clinical characteristics of anti-LRP4 in myasthenia gravis. *J Autoimmun*, 52:139–145.

[198] Beck, G., Yabumoto, T., Baba, K., Sasaki, T., Higuchi, O., Matsuo, H. and Mochizuki, H (2016). Double Seronegative Myasthenia Gravis with Anti-LRP4 Antibodies Presenting with Dropped Head and Acute Respiratory Insufficiency. *H. Intern Med*, 55(22):3361-3363. *Epub*, 15.

[199] McKeon, A., Lennon, V. A., LaChance, D. H., Klein, C. J. and Pittock, S. J. (2013). Striational antibodies in a paraneoplastic context. *Muscle Nerve*, 47(4):585-587.

[200] Ohta, M., Ohta, K., Itoh, N., Kurobe, M., Hayashi, K. and Nishitani, H. (1990). Anti-skeletal muscle antibodies in the sera from myasthenic patients with thymoma: identification of anti-myosin, actomysin, actin, and α-actinin antibodies by a solid-phase radioimmunoassay and a western blotting analysis. *Clinica Chimica Acta*, 187(3): 255–264.

[201] Aarli, J. A., Lefvert, A. K. and Tonder, O. (1981). Thymoma specific antibodies in sera from patients with myasthenia gravis demonstrated

by indirect haemagglutination. *Journal of Neuroimmunology*, 1 (4): 421–427,

[202] Vernino, S. Lennon, V. A. (2004). Autoantibody profiles and neurological correlations of thymoma. *Clin Cancer Res*, 10: 7270–7275.

[203] Choi Decroos, E., Hobson-Webb, L. D., Juel, V. C., Massey, J. M. and Sanders, D. B. (2014). Do acetylcholine receptor and striated muscle antibodies predict the presence of thymoma in patients with myasthenia gravis? *Muscle Nerve*, 49(1): 30-34.

[204] McKeon, A., Lennon, V. A., LaChance, D. H., Klein, C. J. and Pittock, S. J. (2013). Striational antibodies in a paraneoplastic context. *Muscle Nerve*, 47(4):585-587.

[205] Suzuki, S., Utsugisawa, K., Nagane, Y., Satoh, T., Terayama, Y., Suzuki, N. and Kuwana, M. (2007). Classification of myasthenia gravis based on autoantibody status. *Archives of Neurology*, 64(8): 1121–1124,

[206] Romi, F., Skeie, G. O., Gilhus, N. E. and Aarli, J. A. (2005). Striational antibodies in myasthenia gravis: reactivity and possible clinical significance. *Archives of Neurology*, 62(3): 442–446.

[207] Mygland, A. Vincent, A., Newsom-Davis, J., Kaminski, H., Zorzato, F., Agius, M., Gilhus, N. E., Aarli, J. A. (2000). Autoantibodies in thymoma associated myasthenia gravis with myositis or neuromyotonia. *Archives of Neurology*, 57(4): 527–531.

[208] Suzuki, S., Utsugisawa, K., Yoshikawa, H., Motomura, M., Matsubara, S., Yokoyama, K., Nagane, Y., Maruta, T., Satoh, T., Sato, H., Kuwana, M. and Suzuki, N. (2009). Autoimmune targets of heart and skeletal muscles in myasthenia gravis. *Archives of Neurology*, 66(11): 1334–1338,

[209] Evoli, A., Minisci, C., Di Schino, C., Marsili, F., Punzi. C., Batocchi, A. P., Tonali, P. A., Doglietto, G. B., Granone, P., Trodella, L., Cassano, A. and Lauriola, L. (2002). Thymoma in patients with MG: characteristics and long-term outcome. *Neurology*, 59(12): 1844–1850.

[210] Iwasa, K., Kato-Motozaki, Y., Furukawa, Y., Maruta, T., Ishida, C., Yoshikawa, H., Yamada, M. (2010). Up-regulation of MHC class I and class II in the skeletal muscles of myasthenia gravis. *J Neuroimmunol*, 225(1-2): 171-174.

[211] Zamecnik, J., Vesely, D., Jakubicka, B., Simkova, L., Pitha, J., Schutzner, J., Mazanec, R., Vogel, H. (2007). Muscle lymphocytic infiltrates in thymoma-associated myasthenia gravis are phenotypically different from those in polymyositis. *Neuromuscul Disord*, 17(11-12): 935-42.

[212] Suzuki, S., Utsugisawa, K., Nagane, Y., Satoh, T. Kuwana, M. and Suzuki, N. (2011). Clinical and immunological differences between early and late-onset myasthenia gravis in Japan. *Journal of Neuroimmunology*, 230:148–152,

[213] Romi, F., Kristoffersen, E. K., Aarli, J. A. and Gilhus, N. E. (2005). The role of complement in myasthenia gravis: serological evidence of complement consumption *in vivo*. *Journal of Neuroimmunology*, 158, (1-2): 191–194.

[214] Skeie, G. O., Bentsen, P. T., Freiburg, A., Aarli, J. A. and Gilhus, N. E. (1998). Cell-mediated immune response against titin in myasthenia gravis: evidence for the involvement of Th1 and Th2 cells. *Scandinavian Journal of Immunology*, 47(1): 76–81.

[215] Ohno, K., Engel, A. G., Shen, X. M., Selcen, D., Brengman, J., Harper, C. M., Tsujino, A. and Milone, M. (2002). Rapsyn mutations in humans cause endplate acetylcholine-receptor deficiency and myasthenic syndrome. *Am J Hum Genet*, 70(4): 875-885.

[216] Skeie, G. O. and Romi, F. (2008). Paraneoplastic myasthenia gravis: immunological and clinical aspects. *Eur J Neurol*; 15(10): 1029–1033.

[217] Agius, M. A., Zhu, S., Aarli, J. A. (1998). Antirapsyn antibodies occur commonly in patients with lupus. *Ann N Y Acad Sci*, 13; 841: 525-526.

[218] Romi, F., Skeie, G. O., Vedeler, C., Aarli, J. A., Zorzato, F. and Gilhus, N. E. (2000). Complement activation by titin and ryanodine receptor autoantibodies in myasthenia gravis. A study of IgG

subclasses and clinical correlations. *Journal of Neuroimmunology,* 111(1-2): 169–176.

[219] Gautel, M., Lakey, A., Barlow, D. P., Holmes, Z., Scales, S., Leonard, K., Labeit, S., Mygland, A., Gilhus, N. E. and Aarli, J. A. (1993). Titin antibodies in myasthenia gravis: identification of a major immunogenic region of titin. *Neurology,* 43(8): 1581–1585.

[220] F. E. Somnier, G. O. Skeie, J. A. Aarli and Trojaborg, W. (1999). EMG evidence of myopathy and the occurrence of titin autoantibodies in patients with myasthenia gravis. *European Journal of Neurology,* 6(5): 555–563.

[221] Szczudlik, P., Szyluk, B., Lipowska, M., Ryniewicz, B., Kubiszewska, J., Dutkiewicz, M., Gilhus, N. E. and Kostera-Pruszczyk, A. (2014). Antititin antibody in early- and late-onset myasthenia gravis. *Acta Neurol Scand,* 130(4): 229.

[222] Marx, A., Wilisch, A., Schultz, A. Greiner, A., Magi, B., Pallini, V., Schalke, B., Toyka, K., Nix, W., Kirchner, T. and Müller-Hermelink, H. K. (1996). Expression of neurofilaments and of a titin epitope in thymic epithelial tumors. Implications for the pathogenesis of Myasthenia gravis. *American Journal of Pathology,* 148(6): 1839–1850.

[223] Romi, F., Bo, L., Skeie, G. O., Myking, A., Aarli, J. A. and Gilhus, N. E. (2002). Titin and ryanodine receptor epitopes are expressed in cortical thymoma along with costimulatory molecules. *Journal of Neuroimmunology,* 128(1-2): 82–89,

[224] Williams, C. L., Hay, J. E., Huiatt, T. W. and Lennon, V. A. (1992). Paraneoplastic IgG striational autoantibodies produced by clonal thymic B cells and in serum of patients with myasthenia gravis and thymoma react with titin. *Laboratory Investigation,* 66(3): 331–336.

[225] Aarli, J. A. (2008). Myasthenia gravis in the elderly: is it different? *Annals of the New York Academy of Sciences,* 1132: 238– 243.

[226] Romi, F., Skeie, G. O., Gilhus, N. E. and Aarli, J. A. (2005). Striational antibodies in myasthenia gravis: reactivity and possible clinical significance. *Archives of Neurology,* 62(3): 442–446.

[227] Giraud, M., Beaurain, G., Yamamoto A. M., Eymard, B., Tranchant, C., Gajdos, P. and Garchon, H. J. (2001). Linkage of HLA to myasthenia gravis and genetic heterogeneity depending on anti-titin antibodies. *Neurology*, 57(9): 1555–1560.

[228] Saruhan-Direskeneli, G., Kilic, A., Parman, Y., Serdaroglu, P., Deymeer, F. (2006). HLA-DQ polymorphism in Turkish patients with myasthenia gravis. *Hum Immunol*, 67:352-358.

[229] Stergiou, C., Lazaridis, K., Zouvelou V., Tzartos J., Mantegazza R., Antozzi C., Andreetta F., Evoli A., Deymeer F., Saruhan-Direskeneli G., Durmus H., Brenner T., Vaknin A., Berrih-Aknin S., Behin A., SharsharT., De Baets M., Losen M., Martinez-Martinez P., Kleopa, KA., Zamba-Papanicolaou, E., Kyriakides, T., Kostera-Pruszczyk, A., Szczudlik, P., Szyluk, B., Lavrnic, D., Basta, I., Peric, S., Tallaksen, C., Maniaol, A., Gilhus, N. E., Casasnovas Pons, C., Pitha, J., Jakubíkova, M., Hanisch, F., Bogomolovas, J., Labeit, D., Labeit, S. and Tzartos, S. J. (2016). Titin antibodies in "seronegative" myasthenia gravis. A new role for an old antigen. *J Neuroimmunol*, 15; 292: 108-115.

[230] Nakata, M., Kuwabara, S., Kawaguchi, N., Takahashi, H., Misawa, S., Kanai, K., Tamura, N., Sawai, S., Motomura, M., Shiraishi, H., Takamori, M., Maruta, T., Yoshikawa, H., Hattori, T. (2007). Is excitation contraction coupling impaired in myasthenia gravis? *Clinical Neurophysiology*, 118 (5): 1144–1148.

[231] Tsuda, E., Imai, T., Hozuki T., Yamauchi, R., Saitoh, M., Hisahara, S., Yoshikawa, H., Motomura, M. and Shimohama, S. (2010). Correlation of bite force with excitation-contraction coupling time of the masseter in myasthenia gravis. *Clinical Neurophysiology*, 121(7): 1051–1058

[232] Takamori, M. (2008). Autoantibodies against TRPC3 and ryanodine receptor in myasthenia gravis. *Journal of Neuroimmunology*, 200(1-2): 142–144.

[233] Maruta, T., Yoshikawa, H., Fukasawa, S., Umeshita, S., Inaoka, Y., Edahiro, S., Kado, H., Motozaki, Y., Iwasa, K. and Yamada, M.

(2009). Autoantibody to dihydropyridine receptor in myasthenia gravis. *Journal of Neuroimmunology*, 208(1-2): 125–129.

[234] Romi, F., Aarli, J. A., Gilhus, N. E. (2007). Myasthenia gravis patients with ryanodine receptor antibodies have distinctive clinical features. *Eur J Neurol*, 14(6):617-620.

[235] Mygland A., Kuwajima G., Mikoshiba K., Tysnes O. B., Aarli J. A. and Gilhus, N. E. (1995). Thymomas express epitopes shared by the ryanodine receptor. *Journal of Neuroimmunology*, 62(1), pp. 79–83.

[236] Kusner, L. L., Mygland, A. and Kaminski, H. J. (1998). Ryanodine receptor gene expression thymomas. *Muscle and Nerve*, 21(10): 1299–1303.

[237] Mygland A., TysnesO. B., Matre R., Aarli J. A. and Gilhus, N. E. (1994). Anti-cardiac ryanodine receptor antibodies in thymoma-associated myasthenia gravis. *Autoimmunity*, 17(4): 327–331.

[238] Legay, C. and Mei, L. (2017). Moving forward with the neuromuscular junction. *J Neurochem.* 142 Suppl 2: 59-63.

[239] Yellen, G. (2002). The voltage-gated potassium channels and their relatives. *Nature*, 419: 35–42.

[240] Romi, F., Suzuki, S., Suzuki, N., Petzold, A., Plant, G. T., Gilhus, N. E. (2012). Anti-voltage-gated potassium channel Kv1.4 antibodies in myasthenia gravis. *J Neurol*, 259(7): 1312-1316.

[241] Suzuki, S., Satoh, T., Yasuoka, H., Hamaguchi, Y., Tanaka, K., Kawakami, Y., Suzuki, N., Kuwana, M. (2005). Novel auto-antibodies to a voltage-gated potassium channel Kv1.4 in a severe form of myasthenia gravis. *J Neuroimmunol* 170(1–2): 141–149.

[242] Sato, H., Iwasaki, E., Nogawa, S., Suzuki, S., Amano, T., Fukuuchi, Y., Shimoda, M. and Okada, Y. (2003). A patient with giant cell myocarditis and myositis associated with thymoma and myasthenia gravis. *Clinical Neurology*, 43(8): 496– 499.

[243] Tsugawa, J., Tsuboi, Y., Inoue, H., Suzuki, S. and Yamada, T. (2011). Recurrent syncope due to sick sinus syndrome in a patient with myasthenia gravis associated with thymoma. *Clinical Neurology*, 51 (1): 32–34.

[244] Nakamura, K., Katayama, Y., Kusano, K. F., Haraoka, K., Tani, Y., Nagase, S., Morita, H., Miura, D., Fujimoto, Y., Furukawa, T., Ueda, K., Aizawa, Y., Kimura, A., Kurachi, Y. and Ohe, T. (2007). Anti-KCNH2 Antibody-Induced Long QT Syndrome. Novel Acquired Form of Long QT Syndrome. *Journal of the American College of Cardiology,* 50(18): 1808–1809.

[245] Kumagai, S., Kato, T., Ozaki, A., Hirose. S., Minamino, E., Kimura, Y., Nakane, E., Miyamoto, S., Izumi, T., Haruna, T., Nohara, R. and Inoko, M. (2013). Serial measurements of cardiac troponin I in patients with myasthenia gravis-related cardiomyopathy. *Int J Cardiol.* 168(2): e79-80.

[246] Stavroulakis, G., Papadopoulou. M., Koutroulis, G., Zouvelou, V., Katsavochristos, J., Georgiadis, E., Baltogiannis, C. and Avrampos, G. (2015). Myasthenia gravis. A potential cause of false positively elevated troponin T? Case report and brief review. *Int J Cardiol,* Nov 15;199:40-1.

[247] Romi, F., Gilhus, N. E., Varhaug, J. E., Myking, A. and Aarli, J. A. (2003). Disease severity and outcome in thymoma myasthenia gravis: a long-term observation study. *European Journal of Neurology,* 10(6): 701–706.

[248] Romi, F., Skeie, G. O., Aarli, J. A. and Gilhus, N. E. (2000). The severity of myasthenia gravis correlates with the serum concentration of titin and ryanodine receptor antibodies. *Archives of Neurology,* 57(11): 1596–1600.

[249] Takamori, M., Motomura, M., Kawaguchi, N., Nemoto, Y., Hattori, T., Yoshikawa, H. and Otsuka, K. (2004). Antiryanodine receptor antibodies and FK506 in myasthenia gravis. *Neurology,* 62(10): 1894–1896.

[250] Nagane, Y., Suzuki, S., Suzuki. N., Utsugisawa, K. (2010). Factors associated with response to calcineurin inhibitors in myasthenia gravis. *Muscle and Nerve,* 41(2): 212-218.

[251] Zong, Jin, R. (2013). Structural mechanisms of the agrin-LRP4-MuSK signaling pathway in neuromuscular junction differentiation. *Cell Mol Life Sci,* 70(17): 3077-3088.

[252] Bezakova, G. and Ruegg, M. A. (2003). New insights into the roles of agrin. *Nat Rev Mol Cell Biol*, 4:295–308.

[253] Ferns, M. J., Campanelli, J. T., Hoch, W., Scheller, R. H., Hall, Z. (1993). The ability of agrin to cluster AChRs depends on alternative splicing and on cell surface proteoglycans. *Neuron*, 11: 491–502.

[254] Hoch, W., Campanelli, J. T., Scheller, R. H. (1994). Structural domains of agrin required for clustering of nicotinic acetylcholine receptors. *EMBO J*, 13: 2814–2821.

[255] McMahan, U. J., Horton, S. E., Werle, M. J., Honig, L. S., Kroger, S., Ruegg, M. A., Escher, G. (1992). Agrin isoforms and their role in synaptogenesis. *Curr Opin Cell Biol*, 4: 869–874.

[256] DeChiara, T. M., Bowen, D. C., Valenzuela, D. M., Simmons, M. V., Poueymirou, W. T., Thomas, S., Kinetz, E., Compton, D. L., Rojas, E., Park, J. S., Smith, C., DiStefano, P. S., Glass, D. J., Burden, S. J. and Yancopoulos, G. D. (1996). The receptor tyrosine kinase MuSK is required for neuromuscular junction formation *in vivo*. *Cell*, 85(4): 501–512.

[257] Lin, W., Burgess, R. W., Dominguez, B., Pfaff, S. L., Sanes, J. R. and Lee, K. F. (2001). Distinct roles of nerve and muscle in postsynaptic differentiation of the neuromuscular synapse. *Nature*, 410: 1057–1064.

[258] Yang, X., Arber, S., William, C., Li, L., Tanabe, Y., Jessell, T. M., Birchmeier, C. and Burden, S. J. (2001). Patterning of muscle acetylcholine receptor gene expression in the absence of motor innervation. *Neuron*, 30: 399–410.

[259] Gautam, M., Noakes, P. G., Moscoso, L., Rupp, F., Scheller, R. H., Merlie, J. P., Sanes, J. R. (1996). Defective neuromuscular synaptogenesis in agrin-deficient mutant mice. *Cell*, 85: 525–535.

[260] Ruegg, M. A. and Bixby, J. L. (1998). Agrin orchestrates synaptic differentiation at the vertebrate neuromuscular junction. *Trends Neurosci*, 21: 22–27.

[261] Weatherbee, S. D., Anderson, K. V., Niswander, L. A. (2006). LDL-receptor-related protein 4 is crucial for formation of the neuro-muscular junction. *Development*, 133: 4993–5000.

[262] Wu, H., Lu, Y., Shen, C., Patel, N., Gan, L., Xiong, W. C., Mei, L. (2012). Distinct roles of muscle and motoneuron LRP4 in neuromuscular junction formation. *Neuron* 75: 94–107.

[263] Gasperi, C., Melms, A., Schoser, B., Zhang, Y., Meltoranta, J., Risson, V., Schaeffer, L., Schalke, B. and Kröger, S. (2014). Anti-agrin autoantibodies in myasthenia gravis. *Neurology*, 82, 1976–1983.

[264] Zhang, B., Shen, C., Bealmear, B., Ragheb, S., Xiong, W. C., Lewis, R. A., Lisak, R. P. and Mei, L. (2014). Autoantibodies to agrin in myasthenia gravis patients. *PloS One*, 9, e91816.

[265] Cossins, J., Belaya, K., Zoltowska, K., Koneczny, I., Maxwell, S., Jacobson, L., Leite, M. I., Waters, P., Vincent, A. and Beeson, D. (2012). The search for new antigenic targets in myasthenia gravis. *Ann NY Acad Sci*, 1275: 123–128.

[266] Gasperi, C., Melms, A., Schoser, B., Zhang, Y., Meltoranta, J., Risson, V., Schaeffer, L., Schalke B., Kröger S. Anti-agrin autoantibodies in myasthenia gravis. *Neurology*, (2014). 82(22): 1976–1983.

[267] Zhang, B., Shen. C., Bealmear, B. (2014). Autoantibodies to agrin in myasthenia gravis patients. *PLoS One*, 9(3): e91816.

[268] Peng, H. B., Xie, H., Rossi, S. G., Rotundo, R. L. (1999). Acetylcholinesterase clustering at the neuromuscular junction involves perlecan and dystroglycan. *J Cell Biol*, 145: 911–921.

[269] Kimbell, L. M., Ohno, K., Engel, A. G., Rotundo, R. L. (2004). C-terminal and heparin-binding domains of collagenic tail subunit are both essential for anchoring acetylcholinesterase at the synapse. *J Biol Chem*, 279: 10997–11005.

[270] Cartaud, L., Strochlic, M. Guerra, B. Blanchard, M. Lambergeon, E. Krejci, et al. (2004). MuSK is required for anchoring acetyl-cholinesterase at the neuromuscular junction *J Cell Biol*, 165 pp. 505–515.

[271] Kimbell, L. M., Ohno K., Engel A. G. and Rotundo R. L. (2004). C-terminal and heparin-binding domains of collagenic tail subunit are

both essential for anchoring acetylcholinesterase at the synapse. *J Biol Chem*, 279, 10997–11005.

[272] Ohno, K., Brengman, J., Tsujino, A., Engel, A. G. (1998). Human endplate acetylcholinesterase deficiency caused by mutations in the collagen-like tail subunit (ColQ) of the asymmetric enzyme. *Proc Natl Acad Sci U S A*, 95: 9654–9659.

[273] Rotundo, R. L., Rossi, S. G., Kimbell, L. M., Ruiz, C., Marrero, E. (2005). Targeting acetylcholinesterase to the neuromuscular synapse. *Chem Biol Interact*, (157–158): 15–21.

[274] Zoltowska Katarzyna, M., Belaya, K., Leite, M., Patrick, W., Vincent, A. and Beeson, D. (2015). Collagen Q--a potential target for autoantibodies in myasthenia gravis. *J Neurol Sci*, 15; 348(1-2): 241-244.

[275] Kawakami Y. et al. (2011). Anti-MuSK autoantibodies block binding of collagen Q to MuSK. *Neurology* 77, 1819–1826.

[276] Patel, V., Oh, A., Voit, A., Sultatos, L. G., Babu, G. J., Wilson, B. A., Ho, M., McArdle, J. J. (2014). Altered active zones, vesicle pools, nerve terminal conductivity, and morphology during experimental MuSK myasthenia gravis. *PLoS One*, 9(12): e110571.

[277] Phillips, T. M., Manz, H. J., Smith, F. A., Jaffe, H. A., Cohan, S. L. (1981). The detection of anti-cholinesterase antibodies in myasthenia gravis. Ann NY *Acad Sci*, 377: 360–371.

[278] Muller, J. S., Mihaylova, V., Abicht, A., Lochmuller, H. (2007). Congenital myasthenic syndromes: spotlight on genetic defects of neuromuscular transmission. *Expert Rev Mol Med*, 9: 1–20.

[279] Peng, H. B., Xie, H., Dai, Z. (1997). Association of cortactin with developing neuromuscular specializations. *J Neurocytol*, 26: 637–650.

[280] Madhavan, R., Gong, Z. L., Ma, J. J., Chan, A. W. and Peng H. B. (2009). The function of cortactin in the clustering of acetylcholine receptors at the vertebrate neuromuscular junction. *PLoS One*, 4 p. e8478.

[281] Gallardo, E., Martínez-Hernández, E., Titulaer, M. J., Huijbers, M. G., Martínez, M. A., Ramos, A., Querol, L., Díaz-Manera, J., Rojas-

García, R., Hayworth, C. R., Verschuuren, J. J., Balice-Gordon, R., Dalmau, J. and Illa, I. (2014). Cortactin autoantibodies in myasthenia gravis. *Autoimmun Rev*, 13(10):1003-1007.

[282] Labrador-Horrillo, M., Martínez, M. A., Selva-O'Callaghan, A., Trallero-Araguás, E., Grau-Junyent, J. M., Vilardell-Tarrés, M. and Juarez, C. (2014). Identification of a novel myositis-associated antibody directed against cortactin. *Autoimmun Rev*, 13(10):1008-1012.

[283] Cortés-Vicente, E., Gallardo, E., Martínez, M. Á., Díaz-Manera, J., Querol, L., Rojas-García, R. and Illa, I. (2016). Clinical Characteristics of Patients with Double-Seronegative Myasthenia Gravis and Antibodies to Cortactin. *JAMA Neurol*, 73(9):1099-1104.

[284] Ann N Y Acad Sci. 2017 Oct 25. Illa, I., Cortés-Vicente, E., Martínez, M. Á. and Gallardo, E. (2017). Diagnostic utility of cortactin antibodies in myasthenia gravis. *Ann N Y Acad Sci.*

[285] Meager, A., Wadhwa, M., Dilger, P., Bird, C., Thorpe, R., Newsom-Davis, J. and Willcox, N. (2003). Anti-cytokine autoantibodies in autoimmunity: preponderance of neutralizing autoantibodies against interferon-alpha, interferon-omega and interleukin-12 in patients with thymoma and/or myasthenia gravis. *Clin Exp Immunol*, 132(1):128-136.

[286] Buckley, C., Newsom-Davis, J., Willcox, N. and Vincent, A. (2001). Do titin and cytokine antibodies in MG patients predict thymoma or thymoma recurrence? *Neurology*. 13;57(9):1579-1582.

[287] Monstad, S. E., Drivsholm, L., Skeie, G. O., Aarseth, J. H. and Vedeler, C. A. (2008). CRMP5 antibodies in patients with small-cell lung cancer or thymoma. *Cancer Immunol Immunother*, 57(2):227-232.

[288] Zalewski, N. L., Lennon, V. A., Lachance, D. H., Klein, C. J., Pittock, S. J. and Mckeon, A. (2016). P/Q- and N-type calcium-channel antibodies: Oncological, neurological, and serological accompaniments. *Muscle Nerve*, 54(2):220-227.

[289] Leavitt, J. A. (2000). Myasthenia gravis with a paraneoplastic marker. *J Neuroophthalmol*, 20(2):102-105.

[290] Totzeck, A., Mummel, P., Kastrup, O. and Hagenacker, T. (2016). Clinical Features of Neuromuscular Disorders in Patients with N-Type Voltage-Gated Calcium Channel Antibodies. *Eur J Transl Myol*, 15;26(4):5962.

[291] Zekeridou, A., McKeon, A. and Lennon, V. A. (2016). Frequency of Synaptic Autoantibody Accompaniments and Neurological Manifestations of Thymoma. *JAMA Neurol*, 73(7):853-9.

[292] Baekkeskov, S., Aanstoot, H. J., Christgau, S., Reetz, A., Solimena, M., Cascalho, M., Folli, F., Richter-Olesen, H. and De Camilli, P. (1990). Identification of the 64K autoantigen in insulin-dependent diabetes as the GABA-synthesizing enzyme glutamic acid decarboxylase. *Nature*, 347:782.

[293] Kohler, S., Keil, T. O., Swierzy, M., Hoffmann, S., Schaffert, H., Ismail, M., Rückert, J. C., Alexander, T., Hiepe, F., Gross, C., Thiel, A. and Meisel, A. (2013). Disturbed B cell subpopulations and increased plasma cells in myasthenia gravis patients. *J Neuroimmunol*, 15;264(1-2):114-119.

[294] Weiss, J. M., Cufi, P., Bismuth, J., Eymard, B., Fadel, E., Berrih-Aknin, S. and Le Panse, R. (2013). SDF-1/CXCL12 recruits B cells and antigen-presenting cells to the thymus of autoimmune myasthenia gravis patients. *Immunobiology*, 218(3):373-381.

[295] Sun, F., Ladha, S. S., Yang, L., Liu, Q., Shi, S. X., Su, N., Bomprezzi, R. and Shi, F. D. (2014). Interleukin-10 producing-B cells and their association with responsiveness to rituximab in myasthenia gravis. *Muscle Nerve*, 49:487–494.

[296] Yin, W., Allman, W., Ouyang, S., Li, Y., Li, J., Christadoss, P. and Yang, H. (2013). The increased expression of CD21 on AchR specified B cells in patients with myasthenia gravis. *J Neuroimmunol*, 256:49–54.

[297] Wu, H. J. and Bondada, S. (2009). CD72, a coreceptor with both positive and negative effects on B lymphocyte development and function. *J Clin Immunol*, 29:12–21.

[298] Lu, J., Li, J., Zhu, T. Q., Zhang, L., Wang, Y., Tian, F. F. and Yang, H. (2013). Modulation of B cell regulatory molecules CD22 and

CD72 in myasthenia gravis and multiple sclerosis. *Inflammation*, 36(3):521-528.

[299] Kusner, L. L., Ciesielski, M. J., Marx, A., Kaminski, H. J. and Fenstermaker, R. A. (2014). Survivin as a potential mediator to support autoreactive cell survival in myasthenia gravis: a human and animal model study. *PLoS One*, 22;9(7):e102231.

[300] Yi, J. S., Russo, M. A., Massey, J. M., Juel, V., Hobson-Webb, L. D., Gable, K., Raja, S. M., Balderson, K., Weinhold, K. J. and Guptill, J. T. (2017). B10 Cell Frequencies and Suppressive Capacity in Myasthenia Gravis Are Associated with Disease Severity. *Front Neurol*, 10;8:34.

[301] Mizoguchi, A., Mizoguchi, E., Smith, R. N., Preffer, F. I. and Bhan, A. K. (1997). Suppressive role of B cells in chronic colitis of T cell receptor alpha mutant mice. *J Exp Med*, 186: 1749–1756.

[302] Fillatreau, S., Sweenie, C. H., McGeachy, M. J., Gray, D. and Anderton. S. M. (2002). B cells regulate autoimmunity by provision of IL-10. *Nat Immunol,* 3: 944–950.

[303] Mauri, C., Gray, D., Mushtaq, N. and Londei, M. (2003). Prevention of arthritis by interleukin 10-producing B cells. *J Exp Med,* 197: 489–501.

[304] Matsushita, T., Yanaba, K., Bouaziz, J. D., Fujimoto, M. and Tedder, T. F. (2008). Regulatory B cells inhibit EAE initiation in mice while other B cells promote disease progression. *J Clin Invest,* 118: 3420–3430.

[305] Watanabe, R., Ishiura, N., Nakashima, H., Kuwano, Y., Okochi, H., Tamaki, K., Sato, S., Tedder T. F. and Fujimoto, M. (2010). Regulatory B cells (B10 cells) have a suppressive role in murine lupus: CD19 and B10 cell deficiency exacerbates systemic autoimmunity. *J Immunol,* 184: 4801–4809.

[306] Carter, N. A., Rosser, E. C. and Mauri, C. (2012). Interleukin-10 produced by B cells is crucial for the suppression of Th17/Th1 responses, induction of T regulatory type 1 cells and reduction of collagen-induced arthritis. *Arthritis Res Ther,* 14: R32.

[307] Le, Huu. D., Matsushita, T., Jin, G., Hamaguchi, Y., Hasegawa, M., Takehara, K., Tedder, T. F. and Fujimoto, M. (2013). Donor-derived regulatory B cells are important for suppression of murine sclerodermatous chronic graft-versus-host disease. *Blood,* 121: 3274–3283.

[308] Duddy, M., Niino, M., Adatia, F., Hebert, S., Freedman, M., Atkins, H., Kim, H. J. and Bar-Or, A. (2007). Distinct effector cytokine profiles of memory and naive human B cell subsets and implication in multiple sclerosis. *J Immunol,* 178: 6092–6099.

[309] Blair, P. A., Norena, L. Y., Flores-Borja, F., Rawlings, D. J., Isenberg, D. A., Ehrenstein, M. R. and Mauri, C. (2010). CD19(+) CD24(hi)CD38(hi) B cells exhibit regulatory capacity in healthy individuals but are functionally impaired in systemic lupus erythematosus patients. *Immunity,* 32: 129–140.

[310] Newell, K. A., Asare, A., Kirk, A. D., Gisler, T. D., Bourcier, K., Suthanthiran, M., Burlingham, W. J., Marks, W. H., Sanz, I., Lechler, R. I., Hernandez-Fuentes, M. P., Turka, L. A., Seyfert-Margolis, V. L. and Immune Tolerance Network ST507 Study Group. (2010). Identification of a B cell signature associated with renal transplant tolerance in humans. *J Clin Invest,* 120: 1836–1847.

[311] Pallier, A., Hillion, S., Danger, R., Giral, M., Racape, M., Degauque, N., Dugast, E., Ashton-Chess, J., Pettré, S., Lozano, J. J., Bataille, R., Devys, A., Cesbron-Gautier, A., Braudeau, C., Larrose, C., Soulillou, J. P. and Brouard, S. (2010). Patients with drug-free long-term graft function display increased numbers of peripheral B cells with a memory and inhibitory phenotype. *Kidney Int,* 78: 503–513.

[312] Guptill, J. T., Yi, J. S., Sanders, D. B., Guidon, A. C., Juel, V. C., Massey, J. M., Howard, J. F. Jr., Scuderi, F., Bartoccioni, E., Evoli, A. and Weinhold, K. J. (2015). Characterization of B cells in muscle-specific kinase antibody myasthenia gravis. *Neurol Neuroimmunol Neuroinflamm,* 2:e77.10.1212/NXI. 000000 0000000077.

[313] Sheng, J. R., Rezania, K. and Soliven, B. (2016). Impaired regulatory B cells in myasthenia gravis. *J Neuroimmunol,* 297: 38–45.

[314] Ding, T., Yan, F., Cao, S. and Ren, X. (2015). Regulatory B cell: new member of immunosuppressive cell club. *Hum Immunol,* 76: 615–621.

[315] Lykken, J. M., Candando, K. M. and Tedder, T. F. (2015). Regulatory B10 cell development and function. *Int Immunol,* 27: 471–477.

[316] Yanaba, K., Bouaziz, J. D., Haas, K. M., Poe, J. C., Fujimoto, M. and Tedder T. F. (2008). A regulatory B cell subset with a unique CD1dhiCD5+ phenotype controls T cell-dependent inflammatory responses. *Immunity,* 28: 639–650.

[317] Iwata, Y., Matsushita, T., Horikawa, M., Dilillo, D. J., Yanaba, K., Venturi, G. M., Szabolcs, P. M., Bernstein, S. H., Magro, C. M., Williams, A. D., Hall, R. P., St Clair, E. W. and Tedder, T. F. (2011). Characterization of a rare IL-10-competent B-cell subset in humans that parallels mouse regulatory B10 cells. *Blood,* 117: 530–541.

[318] Yi, J. S., Du, M. and Zajac, A. J. (2009). A vital role for interleukin-21 in the control of a chronic viral infection. *Science,* 324: 1572–1576.

[319] Iannello, A., Boulassel, M. R., Samarani, S., Debbeche, O., Tremblay, C., Toma, E., Routy, J. P. and Ahmad, A. (2010). Dynamics and consequences of IL-21 production in HIV-infected individuals: a longitudinal and cross-sectional study. *J Immunol,* 184: 114–126.

[320] Yi, J. S., Ingram, J. T. and Zajac, A. J. (2010). IL-21 deficiency influences CD8 T cell quality and recall responses following an acute viral infection. *J Immunol,* 185: 4835–4845.

[321] Chevalier, M. F., Julg, B., Pyo, A., Flanders, M., Ranasinghe, S., Soghoian, D. Z., Kwon, D. S., Rychert, J., Lian, J., Muller, M. I., Cutler, S., McAndrew, E., Jessen, H., Pereyra, F., Rosenberg, E. S., Altfeld, M., Walker, B. D. and Streeck, H. (2011). HIV-1-specific interleukin-21+ CD4+ T cell responses contribute to durable viral control through the modulation of HIV-specific CD8+ T cell function. *J Virol,* 85: 733–741.

[322] Spolski R. and Leonard, W. J. (2014). Interleukin-21: a double-edged sword with therapeutic potential. *Nat Rev Drug Discov,* 13: 379–395.

[323] Wang, R. X., Yu, C. R., Dambuza, I. M., Mahdi, R. M., Dolinska, M. B., Sergeev, Y. V., Wingfield, P. T., Kim, S. H. and Egwuagu, C. E. (2014). Interleukin-35 induces regulatory B cells that suppress autoimmune disease. *Nat Med,* 20: 633–641.

[324] Sheng, J. R., Quan, S. and Soliven, B. (2014). CD1d(hi)CD5+ B cells expanded by GM-CSF *in vivo* suppress experimental autoimmune myasthenia gravis. *J Immunol,* 193: 2669–2677.

[325] Wong, S. H., Barlow, J. L., Nabarro, S., Fallon, P. G. and McKenzie, A. N. (2010). Tim-1 is induced on germinal centre B cells through B-cell receptor signaling but is not essential for the germinal centre response. *Immunology,* 131: 77–88.

[326] Ding, Q., Yeung, M., Camirand, G., Zeng, Q., Akiba, H., Yagita, H., Chalasani, G., Sayegh, M. H., Najafian, N. and Rothstein, D. M. (2011). Regulatory B cells are identified by expression of TIM-1 and can be induced through TIM-1 ligation to promote tolerance in mice. *J Clin Invest,* 121: 3645–3656.

[327] Xiao, S., Brooks, C. R., Sobel, R. A. and Kuchroo, V. K. (2015). Tim-1 is essential for induction and maintenance of IL-10 in regulatory B cells and their regulation of tissue inflammation. *J Immunol,* 194 (4): 1602–1608.

[328] Yeung, M. Y., Ding, Q., Brooks, C. R., Xiao, S., Workman, C. J., Vignali, D. A., Ueno, T., Padera, R. F., Kuchroo, V. K., Najafian, N. and Rothstein, D. M. (2015). TIM-1 signaling is required for maintenance and induction of regulatory B cells. *Am J Transpl,* 15 (4): 942–953.

[329] Zhang, Y., Zhang, X., Xia, Y., Jia, X., Li, H., Zhang, Y., Shao, Z., Xin, N., Guo, M., Chen, J., Zheng, S., Wang, Y., Fu, L., Xiao, C., Geng, D., Liu, Y., Cui, G., Dong, R., Huang, X. and Yu, T. (2016). CD19+ Tim-1+ B cells are decreased and negatively correlated with disease severity in Myasthenia Gravis patients. *Immunol Res,* 64 (5-6): 1216-1224.

[330] Truffault, F., Cohen-Kaminsky, S., Khalil, I., Levasseur, P. and Berrih-Aknin, S. (1997). Altered intrathymic T-cell repertoire in human myasthenia gravis. *Ann. Neurol,* 41: 731-741.

[331] Grunewald, J., Ahlberg, R., Lefvert, A. K., DerSimonian, H., Wigzell, H. and Janson, C. H. (1991). Abnormal T-cell expansion and V-gene usage in myasthenia gravis patients. Scand. *J. Immunol,* 34: 161-168.

[332] Xu, B. Y., Giscombe, R., Soderlund, A., Troye-Blomberg, M., Pirskanen, R. and Lefvert, A. K. (1998). Abnormal T cell receptor V gene usage in myasthenia gravis: prevalence and characterization of expanded T cell populations. *Clin. Exp. Immunol,* 113: 456-464.

[333] Navaneetham, D., Penn, A. S., Howard, J. F., Jr. and Conti-Fine, B. M. (1998). TCR-Vβ usage in the thymus and blood of myasthenia gravis patients. *J. Autoimm,* 11: 621-633.

[334] Infante, A. J., Billargeon, J., Kraig, E., Lott, L., Jackson, C., Haemmerling, G. J., Raju, R. and David, C. (2003). Evidence of a diverse T cell receptor repertoire for acetylcholine receptor, the autoantigen of myasthenia gravis. *J. Autoimmun,* 21: 167-174.

[335] Matsumoto, Y., Matsuo, H., Sakuma, H., Park, I. K., Tsukada, Y., Kohyama, K., Kondo, T., Kotorii, S. and Shibuya, N. (2006). CDR3 spectratyping analysis of the TCR repertoire in myasthenia gravis. *J Immunol,* 176 (8): 5100-5107.

[336] Tackenberg, B., Schlegel, K., Happel, M., Eienbröker, C., Gellert, K., Oertel, W. H., Meager, A., Willcox, N. and Sommer, N. (2009). Expanded TCR Vbeta subsets of CD8(+) T-cells in late-onset myasthenia gravis: novel parallels with thymoma patients. *J Neuroimmunol,* 216 (1-2): 85-91.

[337] Smith, K. M., Pottage, L., Thomas, E. R., Leishman, A. J., Doig, T. N., Xu, D., Liew, F. Y. and Garside, P. (2000). Th1 and Th2 CD4+ T cells provide help for B cell clonal expansion and antibody synthesis in a similar manner *in vivo. J Immunol,* 165: 3136–3144.

[338] Acosta-Rodriguez, E. V., Napolitani, G., Lanzavecchia, A. and Sallusto, F. (2007). Interleukins 1beta and 6 but not transforming growth factor-beta are essential for the differentiation of interleukin 17-producing human T helper cells. *Nat Immunol,* 8: 942–949.

[339] Wilson, N. J., Boniface, K., Chan, J. R., McKenzie, B. S., Blumenschein, W. M., Mattson, J. D., Basham, B., Smith, K., Chen,

T., Morel, F., Lecron, J. C., Kastelein, R. A., Cua, D. J., McClanahan, T. K., Bowman, E. P. and de Waal Malefyt, R. (2007). Development, cytokine profile and function of human interleukin 17-producing helper T cells. *Nat Immunol*, 8: 950–957.

[340] Sutton, C. E., Lalor, S. J., Sweeney, C. M., Brereton, C. F., Lavelle, E. C. and Mills, K. H. (2009). Interleukin-1 and IL-23 induce innate IL-17 production from gammadelta T cells, amplifying Th17 responses and autoimmunity. *Immunity*, 31: 331–341.

[341] Barrie, A., Khare, A., Henkel, M., Zhang, Y., Barmada, M. M., Duerr, R. and Ray, A. (2011). Prostaglandin E2 and IL-23 plus IL-1b Differentially Regulate the Th1 / Th17 Immune Response of Human CD161 (+) CD4 (+) Memory T Cells. *Clin Transl Sc*i, 4: 268–273.

[342] Ouyang, W., Kolls, J. K. and Zheng, Y. (2008). The biological functions of T helper 17 cell effector cytokines in inflammation. *Immunity*, 28: 454–467.

[343] Dong, C. (2008). Regulation and pro-inflammatory function of interleukin -17 family cytokine. *Immunol*, 226: 80–86.

[344] Pène, J., Chevalier, S., Preisser, L., Vénéreau, E., Guilleux, M. H., Ghannam, S., Molès, J. P., Danger, Y., Ravon, E., Lesaux, S., Yssel, H. and Gascan, H. (2008). Chronically inflamed human tissues are infiltrated by highly differentiated Th17 lymphocytes. *J Immunol*, 180: 7423–7430.

[345] Duerr, R. H., Taylor, K. D., Brant, S. R., Rioux, J. D., Silverberg, M. S., Daly, M. J., Steinhart, A. H., Abraham, C., Regueiro, M., Griffiths, A., Dassopoulos, T., Bitton, A., Yang, H., Targan, S., Datta, L. W., Kistner, E. O., Schumm, L. P., Lee, A. T., Gregersen, P. K., Barmada, M. M., Rotter, J. I., Nicolae, D. L. and Cho, J. H. (2006). A genome-wide association study identifies IL23R as an inflammatory bowel disease gene. *Science*, 314: 1461–1463.

[346] Langrish, C. L., Chen, Y., Blumenschein, W. M., Mattson, J., Basham, B., Sedgwick, J. D., McClanahan, T., Kastelein, R. A. and Cua, D. J. (2005). IL-23 drives a pathogenic T cell population that induces autoimmune inflammation. *J Exp Med*, 201: 233–240.

[347] Weaver, C. T., Hatton R. D., Mangan, P. R. and Harrington, L. E (2007). IL-17 family cytokines and the expanding diversity of effector T cell lineages. *Annu Rev Immunol*, 25: 821–852.

[348] Masuda, M., Matsumoto, M., Tanaka, S., Nakajima, K., Yamada, N., Ido, N., Ohtsuka, T., Nishida, M., Hirano, T. and Utsumi, H. (2010). Clinical implication of peripheral CD4+CD25+ regulatory T cells and Th17 cells in myasthenia gravis patients. *J Neuroimmunol*, 225 (1-2): 123-131.

[349] Schaffert, H., Pelz, A., Saxena, A., Losen, M., Meisel, A., Thiel, A. and Kohler, S. (2015). IL-17-producing CD4(+) T cells contribute to the loss of B-cell tolerance in experimental autoimmune myasthenia gravis. *Eur J Immunol*, 45 (5):1339-1347.

[350] Wang, W., Milan, M., Ostlie, N., Okita, D., Agarwal, R. K., Caspi, R. R. and Conti-Fine, B. M. (2007). C57BL / 6 mice genetically deficient in IL-12 / IL-23 and IFN-gamma are susceptible to experimental autoimmune myasthenia gravis, suggesting a pathogenic role of non-Th1 cells. *J Immunol*, 178:7072–7080.

[351] Mu, L., Sun, B., Kong, Q., Wang, J., Wang, G., Zhang, S., Wang, D., Liu, Y., Liu, Y., An, H. and Li, H. (2009). Disequilibrium of T helper type 1, 2 and 17 cells and regulatory T cells during the development of experimental autoimmune myasthenia gravis. *Immunology*, 128 (Suppl 1):e826–836.

[352] Masuda, M., Matsumoto, M., Tanaka, S., Nakajima, K., Yamada, N., Ido, N., Ohtsuka, T., Nishida, M., Hirano, T. and Utsumi, H. (2010). Clinical implication of peripheral CD4+CD25+ regulatory T cells and Th17 cells in myasthenia gravis patients. *J Neuroimmunol*, 225 (1-2): 123-131.

[353] Roche, J. C., Capablo, J. L., Larrad, L., Gervas-Arruga, J., Ara, J. R., Sánchez, A. and Alarcia, R. (2011). Increased serum interleukin-17 levels in patients with myasthenia gravis. *Muscle Nerve*, 44: 278–280.

[354] Wang, Z., Wang, W., Chen, Y. and Wei, D. (2012). T helper type 17 cells expand in patients with myasthenia-associated thymoma. *Scand J Immunol*, 76 (1): 54-61.

[355] Zhang, Y., Shao, Z., Zhang, X., Jia, X., Xia, Y., Zhang, Y., Xin, N., Guo, M., Chen, J., Zheng, S., Wang, Y., Fu, L., Dong, R., Xiao, C., Geng, D. and Liu Y. (2015). TIPE2 Play a Negative Role in TLR4-Mediated Autoimmune T Helper 17 Cell Responses in Patients with Myasthenia Gravis. *J Neuroimmune Pharmacol*, 10 (4):635-644.

[356] Zhang, Y., Zhang, Y., Li, H., Jia, X., Zhang, X., Xia, Y., Wang, Y., Fu, L., Xiao, C. and Geng, D. (2017). Increased expression of P2X7 receptor in peripheral blood mononuclear cells correlates with clinical severity and serum levels of Th17-related cytokines in patients with myasthenia gravis. *Clin Neurol Neurosurg*, 157: 88-94.

[357] Arulkumaran, N., Unwin, R. J. and Tam, F. W. (2011). A potential therapeutic role for P2X7 receptor (P2X7R) antagonists in the treatment of inflammatory diseases. *Expert Opin Investig Drugs*, 20 (7): 897–915.

[358] Chen, L. and Brosnan, C. F. (2006). Regulation of immune response by P2X7 receptor. *Crit Rev Immunol*, 26 (6): 499–513.

[359] Lister, M. F., Sharkey, J., Sawatzky, D. A., Hodgkiss, J. P., Davidson, D. J., Rossi, A. G. and Finlayson, K. (2007). The role of the purinergic P2X7 receptor in inflammation. *J Inflamm (Lond)*, 4: 5.

[360] Hashiguchi, M., Kobori, H., Ritprajak, P., Kamimura, Y., Kozono, H. and Azuma, M. (2008). Triggering receptor expressed on myeloid cell-like transcript 2 (TLT-2) is a counter-receptor for B7 H3 and enhances T cell responses. *Proceedings of the National Academy of Sciences of the United States of America*, 105 (30):10495-10500.

[361] Janakiram M., Shah U. A., Liu W., Zhao A., Schoenberg M. P. and Zang X. (2017). The third group of the B7-CD28 immune checkpoint family: HHLA2, TMIGD2, B7x, and B7-H3. *Immunol Rev*, 276 (1): 26-39.

[362] Ni, L. and Dong, C. (2017). New checkpoints in cancer immunotherapy. *Immunol Rev*, 276 (1): 52-65.

[363] Veenstra, R. G., Flynn, R., Kreymborg, K., McDonald-Hyman, C., Saha, A., Taylor, P A., Osborn, M. J., Panoskaltsis-Mortari, A., Schmitt, Graeff, A., Lieberknect, E., Murphy, W J., Serody, J. S., Munn, D. H., Freeman, G. J., Allison, J. P., Mak, T. W., van den

Brink, M., Zeiser, R. and Blazar, B. R. (2015). B7-H3 expression in donor T cells and host cells negatively regulates acute graft-versus-host disease lethality. *Blood*, 125 (21): 3335-3346.

[364] Jiang, J. A., Xue, Q., Liu, C. P., Gao, L., Zhang, G. B., Chen, Y. J., Zhang, X. G. (2012). [Expression of B7-H3 costimulatory molecule in peripheral blood of myasthenia gravis patients]. *Xi Bao Yu Fen Zi Mian Yi Xue Za Zhi*, 28 (8):856-859.

[365] Sakthivel P., Wang X., Gharizadeh B., Giscombe R., Pirskanen R., Nyren P. and Lefvert A. K. (2006). Single-nucleotide polymorphisms in the B7H3 gene are not associated with human autoimmune myasthenia gravis. *J Genet*, 85 (3):217-220.

[366] Xiaoyan, Z., Pirskanen, R., Malmstrom, V. and Lefvert, A. K. (2006). Expression of OX40 (CD134) on CD4+ T-cells from patients with myasthenia gravis. *Clin Exp Immunol*, 143 (1): 110-116.

[367] Kotani, A., Takahashi, A., Koga, H., Morita, R., Fukuyama, H., Ichinohe, T., Ishikawa, T., Hori, T. and Uchiyama, T. (2002). Myasthenia gravis after allogeneic bone marrow transplantation treated with mycophenolate mofetil monitored by peripheral blood OX40+ CD4+ T cells. *Eur J Haematol*, 69 (5-6): 318-20.

[368] Chien, P. J., Yeh, J. H., Shih, C. M., Hsueh, Y. M., Chen, M. C. and Chiu, H. C. (2013). A decrease in the percentage of CD3+ cells is correlated with clinical improvement during plasmapheresis in patients with myasthenia gravis. *Artif Organs*, 37 (2):211-216.

[369] Masuda, M., Tanaka, S., Nakajima, K., Yamada, N., Ido, N., Ohtsuka, T., Nishida, M., Hirano, T. and Utsumi H. (2010). Clinical implications of the type 1/type 2 balance of helper T cells and P-glycoprotein function in peripheral T lymphocytes of myasthenia gravis patients. *Eur J Pharmacol*, 627 (1-3):325-331.

[370] Tanaka, S., Masuda, M., Nakajima, K., Ido, N., Ohtsuka, T., Nishida, M., Utsumi, H. and Hirano, T. (2009). P-glycoprotein function in peripheral T lymphocyte subsets of myasthenia gravis patients: clinical implications and influence of glucocorticoid administration. *Int Immunopharmacol*, 9 (3):284-290.

[371] Fan X., Lin C., Han J., Jiang X., Zhu J. and Jin T. (2015). Follicular Helper CD4+ T Cells in Human Neuroautoimmune Diseases and Their Animal Models. *Mediators Inflamm*, 2015:638968.

[372] Saito, R., Onodera, H., Tago, H., Suzuki, Y., Shimizu, M., Matsumura, Y., Kondo, T. and Itoyama, Y. (2005). Altered expression of chemokine receptor CXCR5 on T cells of myasthenia gravis patients. *Journal Neuroimmunol*, 170 (1-2):172–178.

[373] Luo, C., Li, Y., Liu, W., Feng, H., Wang, H., Huang, X., Qiu, L and Ouyang, J. (2013). Expansion of circulating counterparts of follicular helper T cells in patients with myasthenia gravis. *Journal Neuroimmunol*, 256, (1-2), 55–61.

[374] Shiao, Y. M., Lee, C. C., Hsu, Y. H., Huang, S. F., Lin, C. Y., Li, L. H., Fann, C. S., Tsai, C. Y., Tsai, S. F. and Chiu HC. (2010). Ectopic and high CXCL13 chemokine expression in myasthenia gravis with thymic lymphoid hyperplasia. *Journal Neuroimmunol*, 221, (1-2): 101–106.

[375] Meraouna, A., Cizeron-Clairac, G., Panse, R. L., Bismuth, J., Truffault, F., Tallaksen, C. and Berrih-Aknin, S. (2006). The chemokine CXCL13 is a key molecule in autoimmune myasthenia gravis. *Blood*, 108, 2:432–440.

[376] Xin, N., Fu, L., Shao, Z., Guo, M., Zhang, X., Zhang, Y., Dou, C., Zheng, S., Shen, X., Yao Y., Wang, J., Wang, J., Cui, G., Liu, Y., Geng, D., Xiao, C., Zhang, Z. and Dong R. (2014). RNA interference targeting Bcl-6 ameliorates experimental autoimmune myasthenia gravis in mice. *Moll Cell Neurosci*, 58: 85–94.

[377] Thiruppathi, M., Rowin, J., Li Jiang, Q., Sheng, J. R., Prabhakar, B. S. and Meriggioli, M. N. (2012). Functional defect in regulatory T cells in myasthenia gravis. *Ann N Y Acad Sci,* 1274:68-76.

[378] Hori, S., Nomura, T. and Sakaguchi, S. (2003). Control of regulatory T cell development by the transcription factor FOXP3. *Science*, 299: 1057–1061.

[379] Malek, T. R., Yu, A., Vincek, V., Scibelli, P. and Kong L. (2002). CD4 regulatory T cells prevent lethal autoimmunity in IL-2Rbeta-

deficient mice. Implications for the nonredundant function of IL-2. *Immunity*, 17(2): 167-178.

[380] Zorn, E., Nelson, E. A., Mohseni, M., Porcheray, F., Kim, H., Litsa, D., Bellucci, R., Raderschall, E., Canning, C., Soiffer, R J., Frank, D. A. and Ritz, J. (2006). IL-2 regulates FOXP3 expression in human CD4+CD25+ regulatory T cells through a STAT-dependent mechanism and induces the expansion of these cells *in vivo*. *Blood*, 108 (5):1571-1579.

[381] Liao, F. H., Shui, J. W., Hsing, E. W., Hsiao, W. Y., Lin, Y. C., Chan, Y. C., Tan, T. H. and Huang, C. Y. (2014). Protein phosphatase 4 is an essential positive regulator for Treg development, function, and protective gut immunity. *Cell Biosci*, 4:25.

[382] Piccirillo, C. A. and Thornton, A. M. (2004). Cornerstone of peripheral tolerance: naturally occurring CD4+CD25+ regulatory T cells. *Trends Immunol*. 25 (7):374-380.

[383] McHugh, R. S., Whitters, M. J., Piccirillo, C. A., Young, D. A., Shevach, E. M., Collins, M. and Byrne, M. C. (2002). CD4(+) CD25(+) immunoregulatory T cells: gene expression analysis reveals a functional role for the glucocorticoid-induced TNF receptor. *Immunity*, 16 (2):311-323.

[384] Raimondi, G., Shufesky, W. J., Tokita, D., Morelli, A. E. and Thomson, A. W. (2006). Regulated compartmentalization of programmed cell death-1 discriminates CD4+CD25+ resting regulatory T cells from activated T cells. *J Immunol*, 176 (5):2808-2816.

[385] Sharma, M. D., Baban, B., Chandler, P., Hou, D. Y., Singh, N., Yagita, H., Azuma, M., Blazar, B. R., Mellor, A. L. and Munn, D. H. (2007). Plasmacytoid dendritic cells from mouse tumor-draining lymph nodes directly activate mature Tregs via indoleamine 2.3-dioxygenase. *J Clin Invest*, 117 (9): 2570-2582.

[386] Sakthivel, P., Ramanujam, R., Wang, X. B., Pirskanen, R. and Lefvert, A. K. (2008). Programmed Death-1: from gene to protein in autoimmune human myasthenia gravis. *J Neuroimmunol*, 193 (1-2):149-155.

[387] Singh, A. and Kamen DL. (2012). Potential benefits of vitamin D for patients with systemic lupus erythematosus. *Dermatoendocrinol*, 4 (2):146-151.

[388] Urry, Z., Chambers, E. S., Xystrakis, E., Dimeloe, S., Richards, D. F., Gabryšová, L., Christensen, J., Gupta, A., Saglani, S., Bush, A., O'Garra, A., Brown, Z. and Hawrylowicz, C. M. (2012). The role of 1α,25-dihydroxyvitamin D3 and cytokines in the promotion of distinct Foxp3+ and IL-10+ CD4+ T cells. *Eur J Immunol*, 42 (10): 2697-2708.

[389] Askmark, H., Haggård, L., Nygren, I. and Punga, A. R. (2012). Vitamin D deficiency in patients with myasthenia gravis and improvement of fatigue after supplementation of vitamin D3: a pilot study. *Eur J Neurol*, 19 (12): 1554-1560.

[390] Balandina, A., Lécart, S., Dartevelle, P., Saoudi, A. and Berrih-Aknin, S. (2005). Functional defect of regulatory CD4(+)CD25+ T cells in the thymus of patients with autoimmune myasthenia gravis. *Blood*, 105 (2): 735-741.

[391] Luther, C., Adamopoulou, E., Stoeckle, C., Brucklacher-Waldert, V., Rosenkranz, D., Stoltze, L., Lauer, S., Poeschel, S., Melms, A. and Tolosa, E. (2009). Prednisolone treatment induces tolerogenic dendritic cells and a regulatory milieu in myasthenia gravis patients. *J Immunol*. 183 (2): 841-848.

[392] Gradolatto, A., Nazzal, D., Truffault, F., Bismuth, J., Fadel, E., Foti, M. and Berrih-Aknin, S. (2014). Both Treg cells and Tconv cells are defective in the Myasthenia gravis thymus: roles of IL-17 and TNF-α. *J Autoimmun*, 52:53-63.

[393] Wang, H. B., Zhang, J. B. and Chui, L. Y. (2008). Identification of correlations between numbers of CD4+ CD25+ Treg cells, levels of sera anti-AChR antibodies and transfer growth factor-beta in patients with myasthenia gravis. *Zhonghua Yi Xue Za Zhi*. 88 (15):1036-1040. [Article in Chinese]

[394] Li, X., Xiao, B. G., Xi, J. Y., Lu, C. Z. and Lu, J. H. (2008). Decrease of CD4(+)CD25(high)Foxp3(+) regulatory T cells and elevation of

CD19(+)BAFF-R(+) B cells and soluble ICAM-1 in myasthenia gravis. *Clin Immunol*, 126 (2): 180-188.

[395] Kakoulidou, M., Wang, X., Zhao, X., Pirskanen, R., Lefvert, A. K. (2007). Soluble costimulatory factors sCD28, sCD80, sCD86 and sCD152 in relation to other markers of immune activation in patients with myasthenia gravis. *J Neuroimmunol*, 185 (1-2): 150-161.

[396] Baecher-Allan, C., Brown, J. A., Freeman, G. J. and Hafler DA. (2001). CD4+CD25 high regulatory cells in human peripheral blood. *J Immunol.* 167 (3): 1245-1253.

[397] Fattorossi, A., Battaglia, A., Buzzonetti, A., Ciaraffa, F., Scambia, G. and Evoli A. (2005). Circulating and thymic CD4 CD25 T regulatory cells in myasthenia gravis: effect of immunosuppressive treatment. *Immunology.* 116 (1):134-41.

[398] He XT, Liu WB, Feng HY, Zhang Y., Huang X., Meng R. and Wu CY (2008). The role of CD4+ CD25+ T cells in the mechanism of myasthenia gravis in children and adults. *Zhonghua Yi Xue Za Zhi*, 88 (45): 3189-3191. [Article in Chinese]

[399] Nishimura, T., Inaba, Y., Nakazawa, Y., Omata, T., Akasaka, M., Shirai, I. and Ichikawa M. (2015). Reduction in peripheral regulatory T cell population in childhood ocular type myasthenia gravis. *Brain Dev*, 37 (8): 808-816.

[400] Jakubíková, M., Piťha, J., Marečková, H., Týblová, M., Nováková, I. and Schutzner J. (2015). Two-year outcome of thymectomy with or without immunosuppressive treatment in nonthymomatous myasthenia gravis and its effect on regulatory T cells. *J Neurol Sci*, 358 (1-2): 101-106.

[401] Wang Z., Chen Y., Xu S., Yang Y., Wei D., Wang W. and Huang X. (2015). Aberrant decrease of microRNA19b regulates TSLP expression and contributes to Th17 cells development in myasthenia gravis related thymomas. *J Neuroimmunol*, 288: 34-39.

[402] Ziegler, S. F., Roan, F., Bell, B. D., Stoklasek, T. A., Kitajima, M. and Han, H. (2013). The biology of thymic stromal lymphopoietin (TSLP). *Adv. Pharmacol*, 66: 129–155.

[403] Spadoni, I., Iliev, D., Rossi, G. and Rescigno, M. (2012). Dendritic cells produce TSLP that limits the differentiation of Th17 cells, fosters Treg development, and protects against colitis. *Mucosal Immunol.* 5 (2): 184–193.

[404] De Monte, L., Reni, M., Tassi, E., Clavenna, D., Papa, I., Recalde, H., Braga, M., Di Carlo, V., Doglioni, C. and Protti, M. P. (2011). Intratumor T helper type 2 cell infiltrate correlates with cancer associated fibroblast thymic stromal lymphpoietin production and reduced survival in pancreatic cancer. *J. Exp. Med*, 208: 469–478.

[405] Olkhanud, P. B., Rochman, Y., Bodogai, M., Malchinkhuu, E., Wejksza, K., Xu, M., Gress R. E. Hesdorffer, C., Leonard, W. J. and Biragyn, A. (2011). Thymic stromal lymphopoietin is a key mediator of breast cancer progression. *J. Immunol*, 186: 5656–5662.

[406] Benatar M., Sanders D. B., Burns T. M., Cutter G. R., Guptill J. T., Baggi F., Kaminski H. J., Mantegazza R., Meriggioli M. N., Quan J., Wolfe G. I. and Task Force on MG Study Design of the Medical Scientific Advisory Board of the Myasthenia Gravis Foundation of America. (2012). Recommendations for myasthenia gravis clinical trials. *Muscle Nerve*, 45 (6): 909-917.

[407] Roche, J. C., Capablo, J. L., Larrad, L., Gervas-Arruga J., Ara, J. R., Sánchez, A. and Alarcia, R. (2011). Increased serum interleukin-17 levels in patients with myasthenia gravis. *Muscle Nerve*, 44: 278–280.

[408] Mu, L., Sun, B., Kong, Q., Wang, J., Wang, G., Zhang, S., Wang, D., Liu, Y., Liu, Y., An, H. and Li, H. (2009). Disequilibrium of T helper type 1, 2 and 17 cells and regulatory T cells during the development of experimental autoimmune myasthenia gravis. *Immunology*, 128: 826–836.

[409] Zheng, S., Dou, C., Xin, N., Wang, J., Wang, J., Li, P., Fu, L., Shen, X., Cui, G., Dong, R., Lu, J. and Zhang, Y. (2013). Expression of interleukin-22 in myasthenia gravis. *Scand J Immunol*, 78 (1): 98-107.

[410] Ben-Ami, E., Miller, A. and Berrih-Aknin, S. (2014). T cells from autoimmune patients display reduced sensitivity to immuno-

regulation by mesenchymal stem cells: role of IL-2. *Autoimmun Rev*, 13 (2):187-196.

[411] Motobayashi, M., Inaba, Y., Nishimura, T., Kobayashi, N., Nakazawa, Y. and Koike, K. (2015). An increase in circulating B cell-activating factor in childhood-onset ocular myasthenia gravis. *Pediatr Neurol*, 52 (4): 404-409.

[412] Hennerici, T., Pollmann, R., Schmidt, T., Seipelt, M., Tackenberg, B., Möbs, C., Ghoreschi, K., Hertl, M. and Eming, R. (2016). Increased Frequency of T Follicular Helper Cells and Elevated Interleukin-27 Plasma Levels in Patients with Pemphigus. *PLoS One*. 11 (2).

[413] Alahgholi-Hajibehzad, M., Durmuş, H., Aysal, F., Gülşen-Parman, Y., Oflazer, P., Deymeer, F. and Saruhan-Direskeneli, G. (2017). The effect of interleukin (IL)-21 and CD4(+) CD25(++) T cells on cytokine production of CD4(+) responder T cells in patients with myasthenia gravis. *Clin Exp Immunol*, 190 (2): 201-207.

[414] Oliveira, L H., França, Jr. M. C., Nucci, A., Oliveira, D. M., Kimura, E. M. and Sonati Mde F. (2008). Haptoglobin study in myasthenia gravis. *Arq Neuropsiquiatr*, 66 (2A): 229-233.

[415] Suzuki, Y., Onodera, H., Tago, H., Saito, R., Ohuchi, M., Shimizu, M., Matsumura, Y., Kondo, T., Yoshie, O. and Itoyama, Y. (2008). Altered expression of Th1-type chemokine receptor CXCR3 on CD4+ T cells in myasthenia gravis patients. *J Neuroimmunol*, 172 (1-2): 166-174.

[416] Feferman, T., Maiti, P. K., Berrih-Aknin, S., Bismuth, J., Bidault, J., Fuchs, S. and Souroujon, M. C. (2005). Overexpression of IFN-induced protein 10 and its receptor CXCR3 in myasthenia gravis. *J Immunol*, 174 (9): 5324-5331.

[417] Lisak, R. P. and Ragheb, S. (2012). The role of B cell-activating factor in autoimmune myasthenia gravis. *Ann N Y Acad Sci*, 1274: 60-67.

[418] Kang, S. Y., Kang, C. H. and Lee, K. H. (2016). B-cell-activating factor is elevated in serum of patients with myasthenia gravis. *Muscle Nerve*, 54 (6): 1030-1033.

[419] Scuderi, F., Alboini, P. E., Bartoccioni, E. and Evoli, A. (2011). BAFF serum levels in myasthenia gravis: effects of therapy. *J Neurol*, 258 (12): 2284-2285.

[420] Wang, X. B., Kakoulidou, M., Giscombe, R., Qiu, Q., Huang, D., Pirskanen, R. and Lefvert AK. (2002). Abnormal expression of CTLA-4 by T cells from patients with myasthenia gravis: effect of an AT-rich gene sequence. *J Neuroimmunol*; 130 (1-2): 224-232.

[421] Cheng, C., Wu G., Yeung, S. C., Li, R., Bella, A. E., Pang, J., Zhong, F. T., Luo, H., Jin, Y. and Pan, J. (2009). Serum protein profiles in myasthenia gravis. *Ann Thorac Surg*. 88 (4): 1118-1123.

[422] Vincent, F. B., Saulep-Easton, D., Figgett, W. A., Fairfax, K. A. and Mackay, F. (2013). The BAFF/APRIL system: emerging functions beyond B cell biology and autoimmunity. *Cytokine Growth Factor Rev*, 24 (3): 203-215.

[423] Toft-Hansen, H., Nuttall, R. K., Edwards, D. R. and Owens, T. (2004). Key metalloproteinases are expressed by specific cell types in experimental autoimmune encephalomyelitis. *J Immunol*, 173, 5209–5218.

[424] Pietzsch, J. and Hoppmann, S. (2009). Human S100A12: a novel key player in inflammation? *Amino Acids*, 36, 381–389.

[425] Goyette, J. and Geczy, C. L. (2011). Inflammation-associated S100 proteins: new mechanisms that regulate function. *Amino Acids*, 41, 821–842.

[426] Cohen-Kaminsky S., Delattre, R. M., Devergne, O., Klingel-Schmitt, I., Emilie, D., Galanaud, P., Berrih-Aknin, S. (1993). High IL-6 gene expression and production by cultured human thymic epithelial cells from patients with myasthenia gravis. *Ann N Y Acad Sci*, 21; 681: 97-99.

[427] Freund, V. and Frossard, N. (2004). Expression of nerve growth factor in the airways and its possible role in asthma. *Prog Brain Res*, 146, 335–346.

[428] Chunjie, N., Huijuan, N., Zhao, Y., Jianzhao, W. and Xiaojian, Z. (2015). Disease-specific signature of serum miR-20b and its targets

IL-8 and IL-25, in myasthenia gravis patients. *Eur Cytokine Netw*, 26, 61–66.

[429] Berrih-Aknin, S., Ruhlmann, N., Bismuth, J., Cizeron-Clairac, G., Zelman, E., Shachar, I., Dartevelle, P., de Rosbo, N. K. and Le Panse, R. (2009). CCL21 overexpressed on lymphatic vessels drives thymic hyperplasia in myasthenia. *Ann Neurol*, 66 (4): 521-531.

[430] Calabrese, F., Rossetti, A. C., Racagni, G., Gass, P., Riva, M. A. and Molteni, R. (2014). Brain-derived neurotrophic factor: a bridge between inflammation and neuroplasticity. *Front Cell Neurosci*, 22; 8: 430.

[431] Baggi, F., Ubiali, F., Nava, S., Nessi, V. Andreetta, F., Rigamonti, A., Maggi, L., Mantegazza, R. and Antozzi, C. (2008). Effect of IgG immunoadsorption on serum cytokines in MG and LEMS patients. *J Neuroimmunol*; 201-202: 104-110.

[432] Liu A., Lin H., Liu Y., Cao X., Wang X., Li Z. (2009). Correlation of C3 level with severity of generalized myasthenia gravis. *Muscle Nerve*, 40 (5): 801-808.

[433] Christadoss, P., Tüzün, E., Li, J., Saini, S. S. and Yang, H. (2008). Classical complement pathway in experimental autoimmune myasthenia gravis pathogenesis. *Ann N Y Acad Sci*, 1132: 210-219.

[434] Mathai, A., Sarada, C. and Radhakrishnan, V. V. (2000). Significance of circulating immune complexes in myasthenia gravis. *Indian J Med Res*, 111: 180-183.

[435] Li, J., Qi, H., Tuzun, E., Allman, W., Yilmaz, V., Saini, S. S., Deymeer, F., Saruhan-Direskeneli, G. and Christadoss, P. (2009). Mannose-binding lectin pathway is not involved in myasthenia gravis pathogenesis. *J Neuroimmunol*, 208 (1-2): 40-45.

[436] Farrokhi, M., Dabirzadeh, M., Dastravan, N., Etemadifar, M., Ghadimi, K., Saadatpour, Z. and Rezaei, A. (2016). Mannose-binding Lectin Mediated Complement Pathway in Autoimmune Neurological Disorders. *Iran J Allergy Asthma Immunol*, 15 (3): 251-256.

[437] Filková, M., Haluzík, M., Gay, S., Šenolt, L. (2009). The role of resistin as a regulator of inflammation: Implications for various human pathologies. *Clinical Immunology*, 133 (2): 157–170.

[438] Steppan, C. M., Bailey, S. T., Bhat, S., Brown, E. J., Banerjee, R. R., Wright, C. M. and Patel, L. (2001). The hormone resistin links obesity to diabetes. *Nature,* 409: 307–312.

[439] Patel, L., Buckels, A. C., Kinghorn, I. J., Murdock, P. R., Holbrook, J. D., Plumpton, C., Macphee, C. H. and Smith, S. A. (2003). Resistin is expressed in human macrophages and directly regulated by PPAR gamma activators. *Biochem. Biophys. Res. Commun*, 300: 472–476.

[440] Nagaev, I., Bokarewa, M., Tarkowski, A. and Smith, U. (2006). Human resistin is a systemic immune-derived proinflammatory cytokine targeting both leukocytes and adipocytes. *PLoS One* 1, e31.

[441] Kaser, S., Kaser, A., Sandhofer, A., Ebenbichler, C. F., Tilg H. and Patsch, J. R. (2003). Resistin messenger-RNA expression is increased by proinflammatory cytokines *in vitro. Biochemical and Biophysical Research Communications*, 309 (2): 286–290.

[442] Silswal, N., Singh, A. K., Aruna, B., Mukhopadhyay, S., Ghosh, S. and Ehtesham, N. Z. (2005). Human resistin stimulates the pro-inflammatory cytokines TNF-α and IL-12 in macrophages by NF-κB-dependent pathway. *Biochemical and Biophysical Research Communications*, 334 (4): 1092–1101.

[443] Zhang DQ, Wang R., Li T., Li X., Qi Y., Wang J. and Yang L. (2015). Remarkably increased resistin levels in anti-AChR antibody-positive myasthenia gravis. *Journal of Neuroimmunology*, 283: 7–10.

[444] Walcher, D., Hess, K., Berger, R., Aleksic, M., Heinz, P., Bach, H., Durst, R., Hausauer, A., Hombach, V. and Marx, N. (2010). Resistin: a newly identified chemokine for human CD4-positive lymphocytes. *Cardiovasc. Res*, 85: 167–174.

[445] Verma, S., Li, S. H., Wang, C. H., Fedak, P. W., Li, R. K., Weisel, R. D. and Mickle, D. A. (2003). Resistin promotes endothelial cell activation: further evidence of adipokine-endothelial interaction. *Circulation*, 108 (6): 736–740.

[446] Almehed, K., d'Elia, H. F., Bokarewa, M. and Carlsten, H. (2008). Role of resistin as a marker of inflammation in systemic lupus erythematosus. *Arthritis Res. Ther*, 10, R15.

[447] Filkova, M., Senolt, L. and Vencovsky, J. (2013). The role of resistin in inflammatory myopathies. *Curr. Rheumatol. Rep*, 15, 336.

[448] Migita, K., Maeda, Y., Miyashita, T., Kimura, H., Nakamura, M., Ishibashi, H. and Equchi, K. (2006). The serum levels of resistin in rheumatoid arthritis patients. *Clin. Exp. Rheumatol*, 24: 698–701.

[449] Bokarewa, M., Nagaev, I., Dahlberg, L., Smith, U. and Tarkowski, A. (2005). Resistin, an adipokine with potent proinflammatory properties. *J. Immunol,* 174, 5789–5795.

[450] Forsblad d'Elia, H., Pullerits, R., Carlsten, H. and Bokarewa, M. (2008). Resistin in serum is associated with higher levels of IL-1Ra in post-menopausal women with rheumatoid arthritis. *Rheumatology (Oxford),* 47, 1082–1087.

[451] Schaffler, A., Ehling, A., Neumann, E., Herfarth, H., Tarner, I., Scholmerich, J., Müller-Ladner, U. and Gay, S. (2003). Adipocytokines in synovial fluid. *JAMA*, 290, 1709–1710.

[452] Senolt, L., Housa, D., Vernerova, Z., Jirasek, T., Svobodova, R., Voigl, D. Andorlová, K., Müller Ladner, U., Pavelka, K. and Haluzík, M. (2007). Resistin in rheumatoid arthritis synovial tissue, synovial fluid and serum. *Ann. Rheum. Dis*, 66: 458–463.

[453] Bostrom, E. A., d'Elia, H. F., Dahlgren, U., Simark-Mattsson, C., Hasseus, B., Carlsten, H., Tarkowski, A. and Bokarewa, M. (2008). Salivary resistin reflects local inflammation in Sjogren's syndrome. *J. Rheumatol*, 35, 2005–2011.

[454] Da-Qi Zhang, RongWang, Ting Li, Xin Li, Yuan Qi, JingWang, Li Yang. (2015). Remarkably increased resistin levels in anti-AChR antibody-positive myasthenia gravis. *Journal of Neuroimmunology* 283 7–10.

[455] Yang, L., Maxwell, S., Leite, M. I., Waters, P., Clover, L., Fan, X., Zhang, D., Yang, C., Beeson, D. and Vincent A. (2011). Non-radioactive serological diagnosis of myasthenia gravis and clinical features of patients from Tianjin, China. *J. Neurol. Sc,* 301: 71–76.

[456] Leite, M. I., Jacob, S., Viegas, S., Cossins, J., Clover, L., Morgan, B. P., Beeson, D., Willcox, N. and Vincent, A. (2008). IgG1 antibodies to acetylcholine receptors in 'seronegative' myasthenia gravis. *Brain,* 131: 1940–1952.

[457] Di Simone, N., Di Nicuolo, F., Sanguinetti, M., Castellani, R., D'Asta, M., Caforio, L. and Caruso, A. (2006). Resistin regulates human choriocarcinoma cell invasive behaviour and endothelial cell angiogenic processes. *J. Endocrinol,* 189: 691–699.

[458] Housa, D., Vernerova, Z., Heracek, J., Cechak, P., Rosova, B., Kuncova, J. and Haluzik, M. (2008). Serum resistin levels in benign prostate hyperplasia and non-metastatic prostate cancer: possible role in cancer progression. *Neoplasma,* 55, 442–446.

[459] Karmiris, K. and Koutroubakis, I. E. (2007). Resistin: another rising biomarker in inflammatory bowel disease? *Eur J Gastroenterol Hepatol,* 19 (12): 1035-1037.

[460] Konrad, A., Lehrke, M., Schachinger, V., Seibold, F., Stark, R., Ochsenkühn, T., Parhofer, K. G., Göke, B. and Broedl, U. C. (2007). Resistin is an inflammatory marker of inflammatory bowel disease in humans. *Eur J Gastroenterol Hepatol*; 19 (12): 1070-1074.

[461] Zhao, N., Zhang, D. Q., Zhang, L. J., Yang, L. N., Li, L. M., Qi. Y., Wang, J. and Yang, L. (2017) [Clinical significance of serum resistin in patients with generalized myasthenia gravis]. *Zhonghua Yi Xue Za Zhi.* 97 (27): 2087-2090. [Article in Chinese]

[462] Braz, N. F. T., Rocha, N. P., Vieira, É. L. M., Gomez, R. S., Kakehasi, A. M. and Teixeira, A. L. (2017). Body composition and adipokines plasma levels in patients with myasthenia gravis treated with high cumulative glucocorticoid dose. *J Neurol Sci,* 381:169-175.

[463] Janik, S., Schiefer, A. I., Bekos, C., Hacker, P., Haider, T., Moser, J., Klepetko, W., Müllauer, L., Ankersmit, H. J. and Moser B. (2016). HSP27 and 70 expression in thymic epithelial tumors and benign thymic alterations: diagnostic, prognostic and physiologic implications. *Sci Rep,* 6: 24267.

[464] Patil, S. A., Katyayani, S., Sood, A., Kavitha, A. K., Marimuthu, P. and Taly, A. B. (2013). Possible significance of anti-heat shock protein (HSP-65) antibodies in autoimmune myasthenia gravis. *J Neuroimmunol*, 257 (1-2): 107-109.

[465] Mai, W., Liu, X., Fan, Y., Liu, H., Hong, H. Y., Han. R. and Zhou W. (2012). Up-regulated expression of Fas antigen in peripheral T cell subsets in patients with myasthenia gravis. *Clin Invest Med*, 35 (5): E294.

[466] Mizrachi, K., Aricha, R., Feferman, T., Kela-Madar, N., Mandel, I., Paperna, T., Miller, A., Ben-Nun, A., Berrih-Aknin, S., Souroujon, M. C. and Fuchs, S. (2010). Involvement of phosphodiesterases in autoimmune diseases. *J Neuroimmunol*, 220 (1-2): 43-51.

[467] Wang, Y. Z., Yan, M., Tian, F. F., Zhang, J. M., Liu, Q., Yang, H., Zhou, W. B. and Li, J. (2013). Possible involvement of toll-like receptors in the pathogenesis of myasthenia gravis. *Inflammation*, 36 (1): 121-130.

[468] Sengupta, M., Cheema, A., Kaminski, H. J., Kusner, L. L. and Muscle Study Group. (2014). Serum metabolomic response of myasthenia gravis patients to chronic prednisone treatment. *PLoS One*, 9(7):c102635.

[469] Bye, A., Vettukattil, R., Aspenes, S. T., Giskeødegård, G. F., Gribbestad, I. S., Wisløff, U. and Bathen, T. F. (2012). Serum levels of choline-containing compounds are associated with aerobic fitness level: the HUNT-study. *PLoS One*, 7 (7).

[470] Park, H., Bourla, A. B., Kastner, D. L., Colbert, R. A. and Siegel, R. M. (2012). Lighting the fires within: the cell biology of auto-inflammatory diseases. *Nat Rev Immunol*, 12: 570–580.

[471] Stuerenburg, H. J. (2000). The roles of carnosine in aging of skeletal muscle and in neuromuscular diseases. *Biochemistry (Mosc)*, 65: 862–865.

[472] Yang, D., Weng, Y., Lin, H., Xie, F., Yin, F., Lou, K., Zhou, X., Han, Y., Li, X. and Zhang, X. (2016). Serum uric acid levels in patients with myasthenia gravis are inversely correlated with disability. *Neuroreport*, 27:301–305.

[473] Venkatesham, A., Sharath Babu, P., Vidya Sagar, J. and Krishna, D. (2005). Effect of reactive oxygen species on cholinergic receptor function. *Indian J Pharmacol*, 6: 366–370.

[474] Krishnaswamy, A. and Cooper, E. (2012). Reactive oxygen species inactivate neuronal nicotinic acetylcholine receptors through a highly conserved cysteine near the intracellular mouth of the channel: implications for diseases that involve oxidative stress. *J Physiol*, 590 (1): 39–47.

[475] Bourdon, E. and Blache, D. (2001). The importance of proteins in defense against oxidation. *Antioxid. Redox Signal*, 3, 293–311.

[476] Roche, M., Rondeau, P., Singh, N. R., Tarnus, E. and Bourdon, E. (2008). The antioxidant properties of serum albumin. *FEBS Lett*, 582 (13): 1783-1787.

[477] Anraku M. (2014). [Elucidation of the mechanism responsible for the oxidation of serum albumin and its application in treating oxidative stress-related diseases]. *Yakugaku Zasshi*, 134 (9): 973-979. [Article in Japanese]

[478] Fanali, G., di Masi, A., Trezza, V., Marino, M., Fasano, M. and Ascenzi, P. (2012). Human serum albumin: from bench to bedside. *Mol Aspects Med*, 33 (3): 209-290.

[479] Yeo, E. S., Hwang, J. Y., Park, J. E., Choi, Y. J., Huh, K. B. and Kim, W. Y. (2010). Tumor necrosis factor (TNF-alpha) and C-reactive protein (CRP) are positively associated with the risk of chronic kidney disease in patients with type 2 diabetes. *Yonsei Med J*, 51: 519–525.

[480] Heidari B. (2013). C-reactive protein and other markers of inflammation in hemodialysis patients. *Caspian J Intern Med*, 4 (1): 611-616.

[481] Taylor, S. P. and Taylor, B. T. (2012). Healthcare-associated pneumonia in hemodialysis patients: Clinical outcomes in patients treated with narrow versus broad spectrum antibiotic therapy. *Respirology*, 16 (21, 22): 29-31.

[482] Helal, I., Zerelli, L., Krid, M., ElYounsi, F., Ben, Maiz, H., Zouar, B., Adelmoula, J. and Kheder, A. (2012). Comparison of C-reactive

protein and high-sensitivity C-reactive protein levels in patients on hemodialysis. *Saudi J Kidney Dis Transpl*, 23:477–483.

[483] Honda, H., Qureshi, A. R., Heimbürger, O., Barany, P., Wang, K., Pecoits-Filho, R., Stenvinkel, P. and Lindholm, B. (2006). Serum albumin, C-reactive protein, interleukin 6, and fetuin a as predictors of malnutrition, cardiovascular disease, and mortality in patients with ESRD. *Am J Kidney Dis*, 47: 139–148.

[484] Shurraw S. and Tonelli M. (2006). Statins for treatment of dyslipidemia in chronic kidney disease. *Perit Dial Int*; 26: 523–539.

[485] Chen, H. Y., Chiu, Y. L., Hsu, S. P., Pai, M. F., Lai, C. F., Yang, J. Y., Peng, Y. S., Tsai, T. J. and Wu, K. D. (2010). Elevated C-reactive protein level in hemodialysis patients with moderate/severe uremic pruritus: a potential mediator of high overall mortality. *QJM*, 103: 837–846.

[486] Krane, V., Winkler, K., Drechsler, C., Lilientha, J., März, W., Wanner, C. and German Diabetes and Dialysis Study Investigators. (2009). Association of LDL cholesterol and inflammation with cardiovascular events and mortality in hemodialysis patients with type 2 diabetes mellitus. *Am J Kidney Dis*, 54: 902–911.

[487] Rao, M., Jaber, B. L. and Balakrishnan, V. S. (2006). Inflammatory biomarkers and cardiovascular risk: association or cause and effect? *Semin Dial*, 19: 129–135.

[488] Poon, P. Y., Szeto, C. C., Kwan, B. C., Chow, K. M., Leung, C. B. and Li, P. K. (2012). Relationship between serum levels of tumour necrosis factor-related apoptosis-inducing ligand and the survival of Chinese peritoneal dialysis patients. *Nephrology (Carlton)*, 17: 466–471.

[489] Panichi, V., Rizza, G. M., Paoletti, S., Bigazzi, R., Aloisi, M., Barsotti, G., Rindi, P., Donati, G., Antonelli, A., Panicucci, E., Tripepi, G., Tetta, C., Palla, R. and RISCAVID Study Group. (2008). Chronic inflammation and mortality in haemodialysis: effect of different renal replacement therapies. Results from the RISCAVID study. *Nephrol Dial Transplant*, 23 (7): 2337-2343.

[490] Weng, Y. Y., Yang, D. H., Qian, M. Z., Wei, M. M., Yin, F., Li, J., Li, X., Chen, Y., Ding, Z. N., He, Y. B. and Zhang, X. (2016). Low serum albumin concentrations are associated with disease severity in patients with myasthenia gravis. *Medicine (Baltimore)*, 95(39): e5000.

[491] Yang, D. H., Su, Z. Q., Chen, Y., Chen, Z. B., Ding, Z. N., Weng, Y. Y., Li, J., Li, X., Tong, Q. L., Han, Y. X. and Zhang, X. (2016). [Value of the albumin to globulin ratio in predicting severity and prognosis in myasthenia gravis patients]. *Zhonghua Yi Xue Za Zhi*, 96 (9): 697-701. [Article in Chinese]

[492] Davies, K. J., Sevanian, A., Muakkassah-Kelly, S. F. and Hochstein, P. (1986). Uric acid–iron ion complexes. A new aspect of the antioxidant functions of uric acid. *Biochem J*, 235: 747–754.

[493] Glantzounis, G. K., Tsimoyiannis, E. C., Kappas, A. M. and Galaris, D. A. (2005) Uric acid and oxidative stress. *Curr Pharm Des*, 11: 4145–4151.

[494] Whiteman, M., Ketsawatsakul, U., Halliwell, B. (2002). A reassessment of the peroxynitrite scavenging activity of uric acid. *Ann N Y Acad Sci*; 962: 242–259.

[495] Moccia, M., Lanzillo, R., Costabile, T., Russo, C., Carotenuto, A., Sasso, G., Postiglione, E., De Luca Picione, C., Vastola, M., Maniscalco, G. T., Palladino, R. and Brescia Morra, V. (2015). Uric acid in relapsing-remitting multiple sclerosis: a 2-year longitudinal study. *J Neurol*, 262: 961–967.

[496] Oh, S. I., Baek, S., Park, J. S., Piao, L., Oh K. W. and Kim, S. H. (2015). Prognostic role of serum levels of uric acid in amyotrophic lateral sclerosis. *J Clin Neurol*, 11:376–382.

[497] Peng, F., Zhong, X., Deng, X., Qiu W., Wu A., Long, Y., Hu X., Li Q., Jiang Y. and Dai Yl. (2010). Serum uric acid levels and neuromyelitis optica. *J Neurol*, 257:1021–1026.

[498] Miller, A., Glass-Marmor, L., Abraham, M., Grossman, I., Shapiro, S. and Galboiz, Y. (2004). Bio-markers of disease activity and response to therapy in multiple sclerosis. *Clin Neurol Neurosurg*, 106: 249–254.

[499] von Geldern G. and Mowry EM. (2012). The influence of nutritional factors on the prognosis of multiple sclerosis. *Nat Rev Neurol*, 8:678–689.

[500] Hooper, D. C., Scott, G. S., Zborek, A., Mikheeva, T., Kean, R. B., Koprowski, H. and Spitsin, S. V. (2000). Uric acid, a peroxynitrite scavenger, inhibits CNS inflammation, blood–CNS barrier permeability changes, and tissue damage in a mouse model of multiple sclerosis. *FASEB J*, 14:691–698.

[501] Hooper, D. C., Spitsin, S., Kean, R. B., Champion, J. M., Dickson, G. M., Chaudhry, I. and Koprowski H. (1998). Uric acid, a natural scavenger of peroxynitrite, in experimental allergic encephalomyelitis and multiple sclerosis. *Proc Natl Acad Sci USA*, 95:675–680.

[502] Koprowski, H., Spitsin, S. V. and Hooper, D. C. (2001). Prospects for the treatment of multiple sclerosis by raising serum levels of uric acid, a scavenger of peroxynitrite. *Ann Neurol*, 49:139.

[503] Yang D., Weng Y., Lin H., Xie F., Yin F., Lou K., Zhou X., Han Y., Li X. and Zhang X. (2016). Serum uric acid levels in patients with myasthenia gravis are inversely correlated with disability. *Neuroreport*, 27 (5): 301-305.

[504] Xin, Y., Cai, H., Wu, L. and Cui, Y. (2016). The Effect of Immunonutrition on the Postoperative Complications in Thymoma with Myasthenia Gravis. *Mediators Inflamm*. 8781740.

[505] Avidan, N., Le Panse, R., Berrih-Aknin, S. and Miller, A. (2014). Genetic basis of myasthenia gravis - a comprehensive review. *J Autoimmun*, 52:146-53.

[506] Mamrut, S., Avidan, N., Truffault, F., Staun-Ram, E., Sharshar, T., Eymard, B., Frenkian, M., Pitha, J., de Baets, M., Servais, L., Berrih Aknin, S. and Miller A. (2017). Methylome and transcriptome profiling in Myasthenia Gravis monozygotic twins. *J Autoimmun* 82: 62-73.

[507] Krol, J., Loedige, I. and Filipowicz, W. (2010). The widespread regulation of microRNA biogenesis, function and decay. *Nat. Rev. Genet*, 11: 597–610.

[508] Hwang, H. W. and Mendell J. T. (2006). MicroRNAs in cell proliferation, cell death, and tumorigenesis. *Br. J. Cancer*, 94: 776–780.

[509] Zhang, J., Li, S., Li, L., Li, M., Guo, C., Yao, J. and Mi, S. (2015). Exosome and exosomal microRNA: trafficking, sorting, and function. *Genomics Proteomics Bioinformatics*, 13, pp. 17–24.

[510] Punga, A. R. and Punga, T. (2017). Circulating microRNAs as potential biomarkers in myasthenia gravis patients. *Ann N Y Acad Sci*, Nov 10. doi: 10.1111/nyas.13510.

[511] Amarilyo, G. and La Cava, A. (2012). miRNA in systemic lupus erythematosus. *Clin. Immunol*, 144: 26–31.

[512] Gandhi, R., Healy, B., Gholipour, T., Egorova, S., Musallam, A., Hussain, M. S., Nejad, P., Patel, B., Hei, H., Khoury, S., Quintana, F., Kivisakk, P., Chitnis, T. and Weiner, H. L. (2013). Circulating microRNAs as biomarkers for disease staging in multiple sclerosis. *Ann. Neurol*, 73: 729–740.

[513] Schwarzenbach, H., Nishida, N., Calin G. A. and Pantel, K. (2014). Clinical relevance of circulating cell-free microRNAs in cancer. *Nat. Rev. Clin. Oncol*, 11: 145–156.

[514] R. Gandhi, B. Healy, T. Gholipour, S. Egorova, A. Musallam, M. S. Hussain, et al. (2013). Circulating microRNAs as biomarkers for disease staging in multiple sclerosis. *Ann. Neurol.*, 73, pp. 729–740.

[515] Jiang, L., Cheng, Z., Qiu, S., Que, Z., Bao, W., Jiang, C., Zou, F., Liu, P. and Liu, J. (2012). Altered let-7 expression in Myasthenia gravis and let-7c mediated regulation of IL-10 by directly targeting IL-10 in Jurkat cells. *Int Immunopharmacol*; 14 (2):217-223.

[516] Cheng, Z., Qiu, S., Jiang, L., Zhang, A., Bao, W., Liu, P. and Liu, J. (2013). MiR-320a is downregulated in patients with myasthenia gravis and modulates inflammatory cytokines production by targeting mitogen-activated protein kinase 1. *J Clin Immunol,* 33 (3): 567-576.

[517] Lu, J., Yan, M., Wang, Y., Zhang, J., Yang, H., Tian, F. F., Zhou, W., Zhang, N. and Li, J. (2013). Altered expression of miR-146a in myasthenia gravis. *Neurosci Lett*; 555: 85–90.

[518] Wang, J., Zheng, S., Xin N., Dou, C., Fu, L., Zhang, X., Chen, J., Zhang, Y., Geng, D., Xiao, C., Cui, G., Shen, X., Lu, Y., Wang, J., Dong, R., Qiao, Y. and Zhang Y. (2013). Identification of novel MicroRNA signatures linked to experimental autoimmune myasthenia gravis pathogenesis: downregulated miR-145 promotes pathogenetic Th17 cell response. *J Neuroimmune Pharmacol*, 8: 1287–1302.

[519] Wang, Y. Z., Tian, F. F., Yan, M., Zhang, J. M., Liu, Q., Lu, J. Y., Zhou, W. B., Yang, H. and Li, J. (2014). Delivery of a miR155 inhibitor by anti-CD20 single-chain antibody into B cells reduces the acetylcholine receptor-specific autoantibodies and ameliorates experimental autoimmune myasthenia gravis. *Clin Exp Immunol*, 176: 207–221.

[520] Barzago, C., Lum, J., Cavalcante, P., Srinivasan, K. G., Faggiani, E., Camera, G., Bonanno, S. Andreetta, F., Antozzi, C., Baggi, F., Calogero, R. A., Bernasconi, P., Mantegazza, R., Mori, L. and Zolezzi F. (2016). A novel infection- and inflammation-associated molecular signature in peripheral blood of myasthenia gravis patients. *Immunobiology*, 221 (11): 1227-1236.

[521] Nogales-Gadea, G., Ramos-Fransi, A., Suárez-Calvet, X., Nava, M., Rojas-García, R., Mosquera, J. L., Díaz-Manera, J., Querol, L., Gallardo, E. and Illa, I. (2014). Analysis of serum miRNA profiles of myasthenia gravis patients, *PLoS One* 9 e91927.

[522] Punga, T., Le Panse, R. Andersson, M., Truffault, F., Berrih-Aknin, S. and Punga, A. R. (2014). Circulating miRNAs in myasthenia gravis: miR-150-5p as a new potential biomarker. *Ann. Clin. Transl. Neurol* 1, 49–58.

[523] Ghisi, M., Corradin, A., Basso, K., Frasson, C., Serafin, V., Mukherjee, S., Mussolin, L., Ruggero, K., Bonanno, L., Guffanti, A., De Bellis, G., Gerosa, G., Stellin, G., D'Agostino, D. M., Basso, G., Bronte, V., Indraccolo, S., Amadori, A. and Zanovello, P. (2011). Modulation of microRNA expression in human T-cell development: targeting of NOTCH3 by miR-150. *Blood*, 117: 7053–7062.

[524] Monticelli, S., Ansel, K. M., Xiao, C., Socci, N. D., Krichevsky, A. M., Thai, T. H., Rajewsky, N., Marks, D. S., Sander, C., Rajewsky, K., Rao, A. and Kosik KS. (2005). MicroRNA profiling of the murine hematopoietic system. *Genome Biol*, 6 (8): R71.

[525] Fenoglio, C., Cantoni, C., De Riz, M., Ridolfi, E., Cortini, F., Serpente, M., Villa, C., Comi, C., Monaco, F., Mellesi, L., Valzelli, S., Bresolin, N., Galimberti, D. and Scarpini, E. (2011). Expression and genetic analysis of miRNAs involved in CD4+ cell activation in patients with multiple sclerosis. *Neurosci. Lett*, 504: 9–12.

[526] Garchow, B. G., Bartulos Encinas, O., Leung, Y. T., Tsao, P. Y., Eisenberg, R. A., Caricchio, R., Obad, S., Petri, A., Kauppinen, S. and Kiriakidou M. (2011). Silencing of microRNA-21 *in vivo* ameliorates autoimmune splenomegaly in lupus mice. *EMBO Mol. Med*, 3: 605–615.

[527] Ruan, Q., Wang, T., Kameswaran, V., Wei, Q., Johnson, D. S., Matschinsky, F., Shi, W. and Chen, YH. (2011). The microRNA-21-PDCD4 axis prevents type 1 diabetes by blocking pancreatic beta cell death. *Proc. Natl. Acad. Sci. U. S. A*, 108: 12030–12035.

[528] Stagakis, E., Bertsias, G., Verginis, P., Nakou, M., Hatziapostolou, M., Kritikos, H., Iliopoulos, D. and Boumpas, D. T. (2011). Identification of novel microRNA signatures linked to human lupus disease activity and pathogenesis: miR-21 regulates aberrant T cell responses through regulation of PDCD4 expression. *Ann. Rheum. Dis.* 70:1496–1506.

[529] Smigielska-Czepiel, K., van den Berg, A., Jellema, P., Slezak-Prochazka, I., Maat, H., van den Bos, H., van der Lei, R. J., Kluiver, J., Brouwer, E., Boots, A. M. and Kroesen, B. J. (2013). Dual role of miR-21 in CD4+ T-cells: activation-induced miR-21 supports survival of memory T-cells and regulates CCR7 expression in naive T-cells. *PLoS One*, 8 (10) e76217.

[530] Kim, T. D., Lee, S. U., Yun, S., Sun, H. N., Lee, S. H., Kim, J. W., Kim, H. M., Park, S. K., Lee, C. W., Yoon, S. R., Greenberg, P. D. and Choi, I. (2011). Human microRNA-27a*targets Prf1 and GzmB expression to regulate NK-cell cytotoxicity. *Blood,* 118: 5476–5486.

[531] Westerberg, E., Molin, C. J., Lindblad, I., Emtner, M. and Punga, A. R. (2016). Physical exercise in myasthenia gravis is safe and improves neuromuscular parameters and physical performance-based measures: a pilot study. *Muscle Nerve*, 56: 207–214.

[532] Gandhi R(1), Healy B., Gholipour T., Egorova S., Musallam A., Hussain M. S., Nejad P., Patel B., Hei H., Khoury S., Quintana F., Kivisakk P., Chitnis T. and Weiner HL. (2013). Circulating microRNAs as biomarkers for disease staging in multiple sclerosis. *Ann Neurol*, 73 (6): 729-740.

[533] Wang, S., Tang, Y., Cui, H., Zhao, X., Luo, X., Pan, W., Huang, X. and Shen, N. (2011). Let-7/miR-98 regulate Fas and Fas-mediated apoptosis. *Genes Immun*. 12, 149–154.

[534] Dominguez-Villar, M., Gautron, A. S., de Marcken, M., Keller, M. J. and Hafler, D. A., (2015). TLR7 induces anergy in human CD4(+) T cells. *Nat. Immunol*. 16, 118–128.

[535] Luo, Z., Li, Y., Liu, X., Luo, M., Xu, L., Luo, Y., Xiao, B. and Yang H. (2015). Systems biology of myasthenia gravis, integration of aberrant lncRNA and mRNA expression changes. *BMC Med Genomics*, 8:13.

[536] Ramanujam, R., Pirskanen, R., Ramanujam, S. and Hammarström L. (2011) Utilizing twins concordance rates to infer the predisposition to myasthenia gravis. *Twin Res Hum Genet*, 14 (2): 129-136.

[537] Namba, T., Shapiro, M. S., Brunner, N. G. and Grob, D. (1971). Myasthenia gravis occurring in twins. *J Neurol Neurosurg Psychiatry*, 34 (5): 531-534.

[538] Pirskanen R. (1977). Genetic aspects in myasthenia gravis. A family study of 264 Finnish patients. *Acta Neurol Scand*. 56 (5): 365-388.

[539] Namba T., Brunner N. G., Brown S. B., Muguruma M. and Grob D. (1971). Familial myasthenia gravis. Report of 27 patients in 12 families and review of 164 patients in 73 families. *Arch Neurol*, 25 (1): 49-60.

[540] Landouré, G., Knight, MA., Stanescu, H., Taye, A. A., Shi, Y., Diallo, O., Johnson, J. O., Hernandez, D., Traynor, B. J., Biesecker, L. G; NIH Intramural Sequencing Center, Elkahloun, A., Rinaldi, C.,

Vincent, A., Willcox, N., Kleta, R., Fischbeck, K. H., Burnett, B. G. (2012). A candidate gene for autoimmune myasthenia gravis. *Neurology*, 79 (4): 342-347.

[541] Nakata, R., Motomura, M., Masuda, T., Shiraishi, H., Tokuda, M., Fukuda, T. Ando, T., Yoshimura, T., Tsujihata, M. and Kawakami, A. (2013). Thymus histology and concomitant autoimmune diseases in Japanese patients with muscle-specific receptor tyrosine kinase-antibody-positive myasthenia gravis. *Eur J Neurol*, 20 (9): 1272-1276.

[542] Ramanujam, R., Piehl, F., Pirskanen, R., Gregersen, P. K. and Hammarstrom L. (2011). Concomitant autoimmunity in myasthenia gravis-lack of association with IgA deficiency. *J Neuroimmunol*, 236: 118-122.

[543] Vandiedonck, C., Beaurain, G., Giraud, M., Hue-Beauvais, C., Eymard, B., Tranchant, C., Gajdos, P., Dausset, J. and Garchon, H. J. (2004). Pleiotropic effects of the 8.1 HLA haplotype in patients with autoimmune myasthenia gravis and thymus hyperplasia. *Proc Natl Acad Sci U S A*, 101: 15464-15469.

[544] Giraud M., Beaurain G., Eymard B., Tranchant C., Gajdos P., Garchon HJ. (2004). Genetic control of autoantibody expression in autoimmune myasthenia gravis: role of the self-antigen and of HLA-linked loci. *Genes Immun*, 5: 398-404.

[545] Janer, M., Cowland, A., Picard, J., Campbell, D., Pontarotti, P., Newsom-Davis, J., Bunce, Welsh, K., Demaine, A., Wilson, A. G. and Willcox, N. (1999). A susceptibility region for myasthenia gravis extending into the HLA-class I sector telomeric to HLA-C. *Hum Immunol*, 60: 909-917.

[546] Candore, G., Modica, M. A., Lio, D., Colonna-Romano, G., Listì, F., Grimaldi, M. P., Russo, M., Triolo, G., Accardo-Palumbo, A., Cuccia, M. C. and Caruso, C. (2003). Pathogenesis of autoimmune diseases associated with 8.1 ancestral haplotype: a genetically determined defect of C4 influences immunological parameters of healthy carriers of the haplotype. *Biomed Pharmacother*; 57: 274-277.

[547] Candore, G., Lio, D., Colonna Romano, G. and Caruso C. (2002). Pathogenesis of autoimmune diseases associated with 8.1 ancestral haplotype: effect of multiple gene interactions. *Autoimmun Rev*, 1: 29-35.

[548] Zhu, W. H., Lu, J. H., Lin, J., Xi, J. Y., Lu, J., Luo, S. S., Qiao, K., Xiao, B. G., Lu, C. Z. and Zhao, C. B. (2012). HLA-DQA1*03:02/ DQB1*03:03:02 is strongly associated with susceptibility to childhood-onset ocular myasthenia gravis in Southern Han Chinese. *J Neuroimmunol*, 247: 81-85.

[549] Xie, Y. C., Qu, Y., Sun, L., Li, H. F., Zhang, H., Shi, H. J., Jiang, B., Zhao, Y., Qiao, S. S., Wang, S. H. and Wang, D. X. (2011). Association between HLADRB1 and myasthenia gravis in a northern Han Chinese population. *J Clin Neurosci*, 18: 1524-1527.

[550] Xie, Y. C., Qu, Y., Sun, L., Li, H. F., Zhang, H., Shi, H. J., Jiang, B., Zhao, Y., Qiao, S. S., Wang, S. H. and Wang, D. X. (2012). Late onset myasthenia gravis is associated with HLA DRB1*15:01 in the Norwegian population. *PLoS One*, 7: 36603.

[551] Testi, M., Terracciano, C., Guagnano, A., Testa, G., Marfia, G. A., Pompeo E. Andreani M. and Massa R. (2012). Association of IILA-DQB1 *05:02 and DRB1 *16 alleles with late-onset, nonthymomatous, AChR-Ab-positive myasthenia gravis. *Autoimmune Dis*, 2012: 541760.

[552] Franciotta D., Cuccia M., Dondi E., Piccolo G. and Cosi V. (2001). Polymorphic markers in MHC class II/III region: a study on Italian patients with myasthenia gravis. *J Neurol Sci*, 190 (1-2): 11-6.

[553] Santos, E., Bettencourt, A., da Silva, A. M., Boleixa, D., Lopes, D., Brás, S., Costa, P. P. E., Lopes, C., Gonçalves, G., Leite, M. I. and da Silva B. M. (2017). HLA and age of onset in myasthenia gravis. *Neuromuscul Disord*, 27 (7): 650-654.

[554] García-Ramos, G., Téllez-Zenteno, J. F., Zapata-Zúñiga, M., Yamamoto-Furusho, J. K., Ruiz-Morales, J. A., Villarreal-Garza, C., Vargas-Alarcón, G., Estañol, B., Llorente, L. and Granados, J. (2003). HLA class II genotypes in Mexican Mestizo patients with myasthenia gravis. *Eur J Neurol*, 10 (6): 707–710.

[555] Yang, H., Hao, J., Peng, X., Simard, A. R., Zhang, M., Xie, Y., Wang, S. (2012). The association of HLA-DQA1*0401 and DQB1*0604 with thymomatous myasthenia gravis in northern Chinese patients. *J Neurol Sci*, 312 (1–2): 57–61.

[556] Vandiedonck, C., Raffoux, C., Eymard, B., Tranchant, C., Dulmet, E., Krumeich, S., Gajdos, P. and Garchon, H. J. J. (2009). Association of HLA-A in autoimmune myasthenia gravis with thymoma. *Neuroimmunol*, 210 (1-2): 120-123.

[557] Amdahl, C., Alseth, E. H., Gilhus, N. E., Nakkestad, H. L. and Skeie, G. O. (2007). Polygenic disease associations in thymomatous myasthenia gravis. *Arch Neurol*, 64 (12): 1729-1733.

[558] Lee, Y. H., Rho, Y. H., Choi, S. J., Ji, J. D., Song, G. G., Nath, S. K. and Harley, J. B. (2007). The PTPN22 C1858T functional poly-morphism and autoimmune diseases-a meta-analysis. *Rheumatology (Oxford)*, 46: 49-56.

[559] Cloutier J. F. and Veillette A. (1999). Cooperative inhibition of T-cell antigen receptor signaling by a complex between a kinase and a phosphatase. *J Exp Med*, 189: 111-121.

[560] Bottini, N., Musumeci, L., Alonso, A., Rahmouni, S., Nika, K., Rostamkhani, M., MacMurray, J., Meloni, G. F., Lucarelli, P., Pellecchia, M., Eisenbarth, G. S., Comings, D. and Mustelin, T. (2004). A functional variant of lymphoid tyrosine phosphatase is associated with type I diabetes. *Nat Genet*, 36: 337-338.

[561] Begovich, A. B., Carlton, V. E., Honigberg, L. A., Schrodi, S. J., Chokkalingam, A. P., Alexander, H. C., Ardlie, K. G., Huang, Q., Smith, A. M., Spoerke J. M., Conn M. T., Chang M., Chang, S. Y., Saiki, R. K., Catanese, J. J., Leong, D. U., Garcia, V. E., McAllister, L. B., Jeffery D. A., Lee, A. T., Batliwalla, F., Remmers, E., Criswell, L. A., Seldin, M. F., Kastner, D. L., Amos, C. I., Sninsky, J. J. and Gregersen, P. K. (2004). A missense single-nucleotide polymorphism in a gene encoding a protein tyrosine phosphatase (PTPN22) is associated with rheumatoid arthritis. *Am J Hum Genet,* 75: 330-337.

[562] Greve, B., Hoffmann, P., Illes, Z., Rozsa, C., Berger, K., Weissert, R. and Melms, A. (2009). The autoimmunity-related polymorphism

PTPN22 1858C/T is associated with anti-titin antibody-positive myasthenia gravis. *Hum Immuno,* 70: 540-542.

[563] Chuang W. Y., Ströbel P., Belharazem D., Rieckmann P., Toyka K. V., Nix W., Schalke B., Gold R., Kiefer R., Klinker E., Opitz A., Inoue M., Kuo T. T., Müller Hermelink H. K. and Marx A. (2009). The PTPN22gain-of-function+1858T(+) genotypes correlate with low IL-2 expression in thymomas and predispose to myasthenia gravis. *Genes Immun,* 10 (8): 667-672.

[564] Adrianto, I., Wang, S., Wiley, G. B., Lessard, C. J., Kelly, J. A., Adler, A. J., Glenn, S. B., Williams, A. H., Ziegler, J. T., Comeau, M. E., Marion, M. C., Wakeland, B. E., Liang, C., Kaufman, K. M., Guthridge, J. M., Alarcón-Riquelme, M. E., BIOLUPUS and GENLES Networks, Alarcón, G. S., Anaya, J., M., Bae, S. C., Kim, J. H., Joo, Y. B., Boackle, S. A., Brown, E. E., Petri, M. A., Ramsey-Goldman, R., Reveille, J. D., Vilá, L. M., Criswell, L. A., Edberg, J. C., Freedman, B. I., Gilkeson, G. S., Jacob, C. O., James, J. A., Kamen, D. L., Kimberly, R. P., Martín, J., Merrill, J. T., Niewold, T. B., Pons-Estel, B. A., Scofield, R. H., Stevens, A. M., Tsao, B. P., Vyse, T. J., Langefeld, C. D., Harley, J. B., Wakeland, K., Moser, K. L., Montgomery, C. G, and Gaffney, P M. (2012). Association of two independent functional risk haplotypes in TNIP1 with systemic lupus erythematosus. *Arthritis Rheum,* 64: 3695-3705.

[565] Lessard, C. J., Li, H., Adrianto, I., Ice, J. A., Rasmussen, A., Grundahl, K. M., Kelly, J. A., Dozmorov, M. G., Miceli-Richard, C., Bowman, S., Lester, S., Eriksson, P., Eloranta, M. L., Brun, J. G., Gøransson, L. G., Harboe, E., Guthridge, J. M., Kaufman, K. M., Kvarnström, M., Jazebi, H., Cunninghame, Graham D. S., Grandits, M. E., Nazmul-Hossain, A. N., Patel, K., Adler, A. J., Maier-Moore, J. S., Farris, A. D., Brennan, M. T., Lessard, J. A., Chodosh, J., Gopalakrishnan, R., Hefner, K. S., Houston, G. D., Huang, A. J., Hughes, P. J., Lewis, D. M., Radfar, L., Rohrer, M. D., Stone, D. U., Wren, J. D., Vyse, T. J., Gaffney, P. M., James, J. A., Omdal, R., Wahren-Herlenius, M., Illei, G. G., Witte, T., Jonsson, R., Rischmueller, M., Rönnblom, L., Nordmark, G., Ng, W. F.; UK

Primary Sjögren's Syndrome Registry, Mariette, X., Anaya, J. M., Rhodus, N. L., Segal, B. M., Scofield, R. H., Montgomery, C. G., Harley, J. B. and Sivils, K. L. (2013). Variants at multiple loci implicated in both innate and adaptive immune responses are associated with Sjogren's syndrome. *Nat Genet*, 45: 1284e92.

[566] Ramirez, V. P., Gurevich, I., Aneskievich, B. J. (2012). Emerging roles for TNIP1 in regulating post-receptor signaling. *Cytokine Growth Factor Rev*, 23:109e18.

[567] Oshima, S., Turer, E. E., Callahan, J. A., Chai, S., Advincula, R., Barrera, J., Shifrin, N., Lee, B., Benedict Yen, T. S., Woo, T., Malynn, B. A. and Ma, A. (2009). ABIN-1 is an ubiquitin sensor that restricts cell death and sustains embryonic development. *Nature*, 457:906e9.

[568] Wagner, S., Carpentier, I., Rogov, V., Kreike, M., Ikeda, F., Lohr, F., Wu, C. J., Ashwell, J. D., Dötsch, V., Dikic, I., Beyaert, R. (2008). Ubiquitin binding mediates the NF-kappaB inhibitory potential of ABIN proteins. *Oncogene*, 27:3739e45.

[569] Phillips WD and Vincent A. (2016). Pathogenesis of myasthenia gravis: update on disease types, models, and mechanisms [version 1; referees: 2 approved]. *F1000Research*, 5(F1000 Faculty Rev):1513.

[570] Sun, L., Meng, Y., Xie, Y., Zhang, H., Zhang, Z., Wang, X., Jiang, B., Li, W., Li, Y., Yang, Z. (2014). CTLA4 variants and haplotype contribute genetic susceptibility to myasthenia gravis in northern Chinese population. *PLoS One*, 8: 9(7):e101986.

[571] Schneider, H., Downey, J., Smith, A., Zinselmeyer, B. H., Rush, C., Brewer, J. M., Wei, B., Hogg, N., Garside, P., Rudd, C. E. (2006). Reversal of the TCR stop signal by CTLA-4. *Science*, 313(5795):1972-1975.

[572] Waterhouse, P., Penninger, J. M., Timms, E., Wakeham, A., Shahinian, A., Lee, K. P., Thompson, C. B., Griesser, H., Mak, T. W. (1995). Lymphoproliferative disorders with early lethality in mice deficient in Ctla-4. *Science*, 270(5238):985-988.

[573] Qureshi, O. S., Zheng, Y., Nakamura, K., Attridge, K., Manzotti, C., Schmidt, E. M., Baker, J., Jeffery, L. E., Kaur, S., Briggs, Z., Hou, T.

Z., Futter, C. E. Anderson, G., Walker, L. S., Sansom, D. M. (2011). Trans-endocytosis of CD80 and CD86: a molecular basis for the cell-extrinsic function of CTLA-4. *Science*, 332(6029):600-603.

[574] Dubois, P. C., Trynka, G., Franke, L., Hunt, K. A., Romanos, J., Curtotti, A., Zhernakova, A., Heap, G. A., Adány, R., Aromaa, A., Bardella, M. T., van den Berg, L. H., Bockett, N. A., de la Concha, E. G., Dema, B., Fehrmann, R. S., Fernández-Arquero, M., Fiatal, S., Grandone, E., Green, P. M., Groen, H. J., Gwilliam, R., Houwen, R. H., Hunt, S. E., Kaukinen, K., Kelleher, D., Korponay-Szabo, I., Kurppa, K., MacMathuna, P., Mäki, M., Mazzilli, M. C., McCann, O. T., Mearin, M. L., Mein, C. A., Mirza, M. M., Mistry, V., Mora, B., Morley, K. I., Mulder, C. J., Murray, J. A., Núñez, C., Oosterom, E., Ophoff, R. A., Polanco, I., Peltonen, L., Platteel, M., Rybak, A., Salomaa, V., Schweizer, J. J., Sperandeo, M. P., Tack, G. J., Turner, G., Veldink, J. H., Verbeek, W. H., Weersma, R. K., Wolters, V. M., Urcelay, E., Cukrowska, B., Greco, L., Neuhausen, S. L., McManus, R., Barisani, D., Deloukas, P., Barrett, J. C., Saavalainen, P., Wijmenga, C., van Heel, D. A. (2010). Multiple common variants for celiac disease influencing immune gene expression. *Nat Genet*, 42(4):295 302.

[575] Barrett, J. C., Clayton, D. G., Concannon, P., Akolkar, B., Cooper, J. D., Erlich, H. A., Julier, C., Morahan, G., Nerup, J., Nierras, C., Plagnol, V., Pociot, F., Schuilenburg, H., Smyth, D. J., Stevens, H., Todd, J. A., Walker, N. M., Rich, S. S. (2009). Type 1 Diabetes Genetics Consortium. Genome-wide association study and meta-analysis find that over 40 loci affect risk of type 1 diabetes. *Nat Genet*, 41(6):703-707.

[576] Chu, X., Pan, C. M., Zhao, S. X., Liang, J., Gao, G. Q., Zhang, X. M., Yuan, G. Y., Li, C. G., Xue, L. Q., Shen, M., Liu, W., Xie, F., Yang, S. Y., Wang, H. F., Shi, J. Y., Sun, W. W., Du, W. H., Zuo, C. L., Shi, J. X., Liu, B. L., Guo, C. C., Zhan, M., Gu, Z. H., Zhang, X. N., Sun, F., Wang, Z. Q., Song, Z. Y., Zou, C. Y., Sun, W. H., Guo, T., Cao, H. M., Ma, J. H., Han, B., Li, P., Jiang, H., Huang, Q. H., Liang, L., Liu, L. B., Chen, G., Su, Q., Peng, Y. D., Zhao, J. J., Ning,

G., Chen, Z., Chen, J. L., Chen, S. J., Huang, W., Song, H. D. China Consortium for Genetics of Autoimmune Thyroid Disease. (2011). A genome-wide association study identifies two new risk loci for Graves' disease. *Nat Genet*, 43(9):897-901.

[577] Gregersen, P. K., Amos, C. I., Lee, A. T., Lu, Y., Remmers, E. F., Kastner, D. L., Seldin, M. F., Criswell, L. A., Plenge, R. M., Holers, V. M., Mikuls, T. R., Sokka, T., Moreland, L. W., Bridges, S. L. Jr, Xie, G., Begovich, A. B., Siminovitch, K. A. (2009). REL, encoding a member of the NF-kappaB family of transcription factors, is a newly defined risk locus for rheumatoid arthritis. *Nat Genet*, 41(7):820-823.

[578] McIntosh K. R., Linsley P. S., Bacha P. A. and Drachman D. B. (1998). Immunotherapy of experimental autoimmune myasthenia gravis: selective effects of CTLA4Ig and synergistic combination with an IL2-diphtheria toxin fusion protein. *J Neuroimmunol*, 87(1-2):136-146.

[579] McIntosh K. R., Linsley P. S., Drachman D. B. (1995). Immunosuppression and induction of anergy by CTLA4Ig *in vitro*: effects on cellular and antibody responses of lymphocytes from rats with experimental autoimmune myasthenia gravis. *Cell Immunol*, 166(1):103-112.

[580] Strobel, P., Chuang, W. Y., Chuvpilo, S., Zettl, A., Katzenberger, T., Kalbacher, H., Rieckmann, P., Nix, W., Schalke, B., Gold, R., Müller-Hermelink, H. K., Peterson, P., Marx, A. (2008). Common cellular and diverse genetic basis of thymoma-associated myasthenia gravis: role of MHC class II and AIRE genes and genetic polymorphisms. *Ann N Y Acad Sci*, 1132:143e56.

[581] Wang X. B., Pirskanen R., Giscombe R. and Lefvert A. K. (2008). Two SNPs in the promoter region of the CTLA-4 gene affect binding of transcription factors and are associated with human myasthenia gravis. *J Intern Med*, 263:61e9.

[582] Gu M., Kakoulidou M., Giscombe R., Pirskanen R., Lefvert A.K., Klareskog L., Wang X. (2008). Identification of CTLA-4 isoforms

produced by alternative splicing and their association with myasthenia gravis. *Clin Immunol*, 128:374e81.

[583] Chuang, W. Y., Ströbel, P., Bohlender-Willke, A. L., Rieckmann, P., Nix, W., Schalke, B., Gold, R., Opitz, A., Klinker, E., Inoue, M., Müller-Hermelink, H. K., Saruhan-Direskeneli, G., Bugert, P., Willcox, N. and Marx, A. (2014). Late-onset myasthenia gravis - CTLA4(low) genotype association and low-for-age thymic output of naïve T cells. *J Autoimmun*, 52:122-129.

[584] Fernández-Mestre M., Sánchez K., Balbás O., Gendzekhzadze K., Ogando V., Cabrera M., Layrisse Z. (2009). Influence of CTLA-4 gene polymorphism in autoimmune and infectious diseases. *Hum Immunol*, 70(7):532-5.

[585] Huang, D., Liu, L., Norén, K., Xia, S. Q., Trifunovic, J., Pirskanen, R., Lefvert, A. K. (1998). Genetic association of Ctla-4 to myasthenia gravis with thymoma. *J Neuroimmunol*, 88(1-2):192-8.

[586] Huang D., Pirskanen R., Hjelmström P., Lefvert AK (1998). Polymorphisms in IL-1beta and IL-1 receptor antagonist genes are associated with myasthenia gravis. *J Neuroimmunol*, 81(1-2):76-81.

[587] Anderson, D. M., Maraskovsky, E., Billingsley, W. L., Dougall, W. C., Tometsko, M. E., Roux, E. R., Teepe, M. C., DuBose, R. F., Cosman, D., Galibert, L. (1997). A homologue of the TNF receptor and its ligand enhance T-cell growth and dendritic-cell function. *Nature*, 390(6656):175-179.

[588] Dougall, W. C., Glaccum, M., Charrier, K., Rohrbach, K., Brasel, K., De Smedt, T., Daro, E., Smith, J., Tometsko, M. E., Maliszewski, C. R., Armstrong, A., Shen, V., Bain, S., Cosman, D. Anderson, D., Morrissey, P. J., Peschon, J. J., Schuh, J. (1999). RANK is essential for osteoclast and lymph node development. *Genes Dev*, 13(18):2412-2424.

[589] Kroeger K. M., Carville K. S., Abraham L. J. (1997). The -308 tumor necrosis factor-alpha promoter polymorphism effects transcription. *Mol Immunol*, 34:391e9.

[590] Rodriguez-Carreon, A. A., Zuniga, J., Hernandez-Pacheco, G., Rodriguez-Perez, J. M., Perez-Hernandez, N., Montes de Oca, J. V.,

Cardiel, M. H., Granados, J. and Vargas-Alarcón, G. (2005). Tumor necrosis factor-alpha-308 promoter polymorphism contributes independently to HLA alleles in the severity of rheumatoid arthritis in Mexicans. *J Autoimmun*, 24:63e8.

[591] O'Keefe, G. E., Hybki, D. L. and Munford, R. S. (2002). The G->A single nucleotide polymorphism at the -308 position in the tumor necrosis factor-alpha promoter increases the risk for severe sepsis after trauma. *J Trauma*, 52:817e25.

[592] Elahi, M. M., Asotra, K., Matata, B. M. and Mastana, S. S. (2009). Tumor necrosis factor alpha-308 gene locus promoter polymorphism: an analysis of association with health and disease. *Biochim Biophys Acta*, 1792:163e72.

[593] Hong, Y., Ge, Z., Jing, C., Shi, J., Dong, X., Zhou, F., Wang, M., Zhang, Z. and Gong W. (2013). Functional promoter-308G>A variant in tumor necrosis factor alpha gene is associated with risk and progression of gastric cancer in a Chinese population. *PLoS One*, 8:e50856.

[594] Huang, D. R., Pirskanen, R., Matell, G. and Lefvert, A. K. (1999). Tumour necrosis factor-alpha polymorphism and secretion in myasthenia gravis. *J Neuroimmunol*, 94:165e71.

[595] Skeie, G. O., Pandey, J. P., Aarli, J. A. and Gilhus, N. E. (1999). TNFA and TNFB polymorphisms in myasthenia gravis. *Arch Neurol*, 56:457e61.

[596] Zelano, G., Lino, M. M., Evoli, A., Settesoldi, D., Batocchi, A. P., Torrente, I. and Tonali P. A. (1998). Tumour necrosis factor beta gene polymorphisms in myasthenia gravis. *Eur J Immunogenet*, 25:403e8.

[597] Josefowicz, S. Z., Lu, L. F. and Rudensky, A. Y. (2012). Regulatory T cells: mechanisms of differentiation and function. *Annu Rev Immunol*, 30:531e64.

[598] Zhang, J., Chen, Y., Jia, G., Chen, X., Lu, J., Yang, H., Zhou, W., Xiao, B., Zhang, N. and Li, J. (2013). FOXP3-3279 and IVS9+459 polymorphisms are associated with genetic susceptibility to myasthenia gravis. *Neurosci Lett*, 534:274-8.

[599] Giraud, M., Eymard, B., Tranchant, C., Gajdos, P. and Garchon, H. J. (2004). Association of the gene encoding the delta-subunit of the muscle acetylcholine receptor (CHRND) with acquired autoimmune myasthenia gravis. *Genes Immun*, 5:80-3.

[600] Heckmann, J. M., Morrison, K. E., Emeryk-Szajewska, B., Strugalska, H., Bergoffen, J., Willcox, N. and Newsom-Davis, J. (1996). Human muscle acetylcholine receptor alpha-subunit gene (CHRNA1) association with autoimmune myasthenia gravis in black, mixed ancestry and Caucasian subjects. *J Autoimmun*, 9:175-80.

[601] Wilisch, A., Gutsche, S., Hoffacker, V., Schultz, A., Tzartos, S., Nix, W., Schalke, B., Schneider, C., Müller-Hermelink, H. K. and Marx, A. (1999). Association of acetylcholine receptor alpha-subunit gene expression in mixed thymoma with myasthenia gravis. *Neurology*, 52:1460-6.

[602] Giraud, M., Taubert, R., Vandiedonck, C., Ke, X., Levi-Strauss, M., Pagani, F., Baralle, F. E., Eymard, B., Tranchant, C., Gajdos, P., Vincent, A., Willcox, N., Beeson, D., Kyewski, B. and Garchon, H. J. (2007). An IRF8-binding promoter variant and AIRE control CHRNA1 promiscuous expression in thymus. *Nature*, 448:934e7.

[603] Szabo, S. J., Kim, S. T., Costa, G. L., Zhang, X., Fathman, C. G., Glimcher, L. H. (2000). A novel transcription factor, T-bet, directs Th1 lineage commitment. *Cell*, 100(6):655-69.

[604] Soderquest, K., Hertweck, A., Giambartolomei, C., Henderson, S., Mohamed, R., Goldberg, R., Perucha, E., Franke, L., Herrero, J., Plagnol, V., Jenner, R. G. and Lord, G. M. (2017). Genetic variants alter T-bet binding and gene expression in mucosal inflammatory disease. *PLoS Genet*, 13(2):e1006587.

[605] Liu, R., Hao, J., Dayao, C. S., Shi, F. D. and Campagnolo, D. I. (2009). T-bet deficiency decreases susceptibility to experimental myasthenia gravis. *Exp Neurol*, 220(2):366-373.

[606] Viken, M. K., Sollid, H. D., Joner, G., Dahl-Jørgensen, K., Rønningen, K. S., Undlien, D. E., Flatø, B., Selvaag, A. M., Førre, Ø., Kvien, T. K., Thorsby, E., Melms, A., Tolosa, E. and Lie, B. A.

(2007). Polymorphisms in the cathepsin L2 (CTSL2) gene show association with type 1 diabetes and early-onset myasthenia gravis. *Hum Immunol*, 68(9):748-55.

[607] Pál, Z., Antal, P., Millinghoffer, A., Hullám, G., Pálóczi, K., Tóth, S., Gabius, H. J., Molnár, M. J., Falus, A., Buzás, E. I. (2010). A novel galectin-1 and interleukin 2 receptor β haplotype is associated with autoimmune myasthenia gravis. *J Neuroimmunol*, 229(1-2):107-11.

[608] Yilmaz, V., Tutuncu, Y., Baris Hasbal, N., Parman, Y., Serdaroglu, P., Deymeer, F. and Saruhan-Direskeneli, G. (2007). Polymorphisms of interferon-gamma, interleukin-10, and interleukin-12 genes in myasthenia gravis. *Hum Immunol*, 68: 544-549.

[609] Pal, Z., Varga, Z., Semsei, A., Remenyi, V., Rozsa, C., Falus, A., Illes, Z., Buzás, E. I., Molnar, M. J. (2012). Interleukin-4 receptor alpha polymorphisms in autoimmune myasthenia gravis in a Caucasian population. *Hum Immunol*, 73: 193-195.

[610] Alseth, E. H., Nakkestad H. L., Aarseth J., Gilhus N. E., Skeie G. O. Interleukin-10 promoter polymorphisms in myasthenia gravis. *J Neuroimmunol*, 2009; 210: 63-66.

[611] Zagoriti, Z., Georgitsi, M., Giannakopoulou, O., Ntellos, F., Tzartos, S. J., Patrinos, G. P. and Poulas, K. (2012). Genetics of myasthenia gravis: a case-control association study in the Hellenic population. *Clin Dev Immunol*, 2012:484919.

[612] Tüzün, E., Saini, S. S., Yang, H., Alagappan, D., Higgs, S., Christadoss, P. (2006). Genetic evidence for the involvement of Fcgamma receptor III in experimental autoimmune myasthenia gravis pathogenesis. *J Neuroimmunol*, 174(1-2): 157-167.

[613] van der Pol, W. L., Jansen, M. D., Kuks, J. B., de Baets, M., Leppers-van de Straat, F. G., Wokke, J. H., van de Winkel, J. G., van den Berg, L. H. (2003). Association of the Fc gamma receptor IIA-R/R131 genotype with myasthenia gravis in Dutch patients. *J Neuroimmunol*, 144(1-2): 143-147.

[614] Avidan, N., Le Panse, R., Harbo, H. F., Bernasconi P., Poulas, K., Ginzburg, E., Cavalcante, P., Colleoni, L., Baggi, F., Antozzi, C., Truffault, F., Horn-Saban, S., Pöschel, S., Zagoriti, Z., Maniaol, A.,

Lie, B. A., Bernard, I., Saoudi, A., Illes, Z., Casasnovas Pons, C., Melms, A., Tzartos, S., Willcox, N., Kostera-Pruszczyk, A., Tallaksen, C., Mantegazza, R., Berrih-Aknin, S. and Miller, A. (2014). VAV1 and BAFF, via NFκB pathway, are genetic risk factors for myasthenia gravis. *Ann Clin Transl Neurol,* 1(5):329-339.

[615] Nel, M., Buys, J. M., Rautenbach, R., Mowla, S., Prince, S., Heckmann, J. M. (2016). The African-387 C>T TGFB1 variant is functional and associates with the ophthalmoplegic complication in juvenile myasthenia gravis. *J Hum Genet.* Apr; 61(4):307-316.

[616] Heckmann, J. M., Uwimpuhwe, H., Ballo, R., Kaur, M., Bajic, V., Prince, B. S. (2010). A functional SNP in the regulatory region of the decay-accelerating factor gene associates with extraocular muscle pareses in myasthenia gravis. *Genes Immun,* 11: 1–10.

[617] Auret, J., Abrahams, A., Prince, S., Heckmann J. M. (2014). The effects of prednisone and steroid-sparing agents on decay accelerating factor (CD55) expression: implications in myasthenia gravis. *Neuromuscul Disord,* 24(6): 499-508.

[618] Kaminski, H. J., Kusner, L. L., Richmonds, C., Medof, M. E., Lin, F. (2006). Deficiency of decay accelerating factor and CD59 leads to crisis in experimental myasthenia. *Exp Neurol*, 202(2):287-293.

[619] Na, S. J., Lee, J. H., Kim. S. W., Kim, D. S., Shon, E. H., Park, H. J., Shin, H. Y., Kim, S. M., Choi, Y. C. (2014). Whole-genome analysis in Korean patients with autoimmune myasthenia gravis. *Yonsei Med J,* 55(3): 660-668.

[620] Park, K. H., Jung, J., Lee, J. H., Hong, Y. H. (2016). Blood Transcriptome Profiling in Myasthenia Gravis Patients to Assess Disease Activity: A Pilot RNA-seq Study. *Exp Neurobiol*, 25(1): 40-7.

[621] Xie, Y., Meng, Y., Li, H. F., Hong, Y., Sun, L., Zhu, X., Yue, Y. X., Gao, X., Wang, S., Li, Y., Kusner, L. L. and Kaminski, H. J. (2016). GR gene polymorphism is associated with inter-subject variability in response to glucocorticoids in patients with myasthenia gravis. *Eur J Neurol.* Aug; 23(8): 1372-1379.

[622] Xie, Y., Li, H. F., Sun, L., Kusner, L. L., Wang, S., Meng, Y., Zhang, X., Hong, Y., Gao, X., Li, Y., Kaminski, H. J. (2017). The Role of Osteopontin and Its Gene on Glucocorticoid Response in Myasthenia Gravis. *Front Neurol.* May 31; 8: 230.

[623] Richaud-Patin, Y., Vega-Boada, F., Vidaller, A., Llorente L. (2004). Multidrug resistance-1 (MDR-1) in autoimmune disorders IV. P-glycoprotein overfunction in lymphocytes from myasthenia gravis patients. *Biomed Pharmacother,* 58(5): 320-324.

[624] Tanaka, S., Hirano, T., Saito, T., Wakata, N., Oka, K. (2007). P-glycoprotein function in peripheral blood mononuclear cells of myasthenia gravis patients treated with tacrolimus. *Biol Pharm Bull.* 30(2): 291-296.

[625] Tavakolpour, S., Darvishi, M., Ghasemiadl, M. (2017). Pharmacogenetics: A strategy for personalized medicine for autoimmune diseases. *Clin Genet*, Nov 30.

[626] Chen, D., Lian, F., Yuan, S. et al. (2014). Association of thiopurine methyltransferase status with azathioprine side effects in Chinese patients with systemic lupus erythematosus. *Clinical Rheumatology*, 33: 499-503.

[627] Colleoni, L., Kapetis, D., Maggi, L., Camera, G., Canioni, E., Cavalcante P., Kerlero de Rosbo, N., Baggi, F., Antozzi, C., Confalonieri, P., Mantegazza, R. and Bernasconi, P. (2013). A new thiopurine s-methyltransferase haplotype associated with intolerance to azathioprine. *J Clin Pharmacol*, 53(1): 67-74.

[628] Stocco, G., Pelin, M., Franca, R., De Iudicibus, S., Cuzzoni, E., Favretto, D., Martelossi, S., Ventura, A. and Decorti, G. (2014). Pharmacogenetics of azathioprine in inflammatory bowel disease: a role for glutathione-S-transferase? *World J Gastroenterol.* Apr 7; 20(13): 3534-3541.

[629] Qiu, Q., Huang, J., Lin, Y., Shu, X., Fan, H., Tu, Z., Zhou, Y. and Xiao, C. (2017). Polymorphisms and pharmacogenomics for the toxicity of methotrexate monotherapy in patients with rheumatoid arthritis: A systematic review and meta analysis. *Medicine* (Baltimore), 96(11):e6337.

[630] Pasnoor M., He J., Herbelin L., Dimachkie M., Barohn RJ; Muscle Study Group. (2012). Phase II trial of methotrexate in myasthenia gravis. *Ann N Y Acad Sci*, 1275: 23-28.

[631] Dervieux T., Kremer J., Orentas Lein D., et al. Contribution of common polymorphisms in reduced folate carrier and γ-glutamyl-hydrolase to Methotrexate polyglutamate levels in patients with rheumatoid arthritis. *Pharmacogenetics*. 2004; 14:733–739.

[632] Derivieux, T., Furst, D., Orentas Lein, D. O., Capps, R., Smith, K., Caldwell, J. and Kremer, J. (2005). Pharmacogenetic and metabolite measurements are associated with clinical status in patients with rheumatoid arthritis treated with Methotrexate: results of a multicentered cross sectional observational study. *Ann Rheum Dis*, 64: 1180–1185.

[633] Chen, Y., Zou, K., Sun, J., Yang, Y. and Liu, G. (2017). Are gene polymorphisms related to treatment outcomes of methotrexate in patients with rheumatoid arthritis? A systematic review and meta-analysis. *Pharmacogenomics*, 18(2):175-195.

[634] Zhang, Y. T., Yang, L. P., Shao, H., Li, K. X., Sun, C. H. and Shi, L. W. (2008). ABCB1 polymorphisms may have a minor effect on ciclosporin blood concentrations in myasthenia gravis patients. *Br J Clin Pharmacol*, 66(2): 240-6.

[635] Benatar, M., Sanders, D. B., Burns, T. M., Cutter, G. R., Guptill, J. T., Baggi, F., Kaminski, H. J., Mantegazza, R., Meriggioli, M. N., Quan, J., Wolfe, G. I. (2012). Task Force on MG Study Design of the Medical Scientific Advisory Board of the Myasthenia Gravis Foundation of America. *Muscle Nerve*. Jun; 45(6):909-17. Recommendations for myasthenia gravis clinical trials.

[636] Barnett, C., Grinberg, Y., Ghani, M., Rogaeva, E., Katzberg, H., St George-Hyslop, P. and Bril, V. (2012). Fcγ receptor polymorphisms do not predict response to intravenous immunoglobulin in myasthenia gravis. *J Clin Neuromuscul Dis*, Sep; 14(1): 1-6.

[637] Mays, J. A. and · Butts, C. L. B. (2011). Intercommunication between the Neuroendocrine and Immune Systems: Focus on Myasthenia Gravis. *Neuroimmunomodulation,* 18: 320–327.

[638] Nancy, P. and Berrih-Aknin, S. (2005). Differential Estrogen Receptor Expression in Autoimmune Myasthenia Gravis. *Endocrinology,* 146 (5): 2345-2353.

[639] Delpy, L., Douin-Echinard. V., Garidou, L., Bruand, C., Saoudi, A. and Guéry, J. C. (2005). Estrogen Enhances Susceptibility to Experimental Autoimmune Myasthenia Gravis by Promoting Type 1-Polarized Immune Responses. *J Immunol,* 175 (8): 5050-5057.

[640] Kaur, M., Schmeier, S., MacPherson, C. R., Hofmann, O., Hide, W. A., Taylor, S., Willcox, N. and Bajic, V. B. (2008). Prioritizing genes of potential relevance to diseases affected by sex hormones: an example of myasthenia gravis. *BMC Genomics,* 9: 481.

[641] Janer, M., Cowland, A., Picard, J., Campbell, D., Pontarotti, P., Newsom-Davis, J., Bunce, M., Welsh, K., Demaine, A., Wilson, A. G. and Willcox, N. (1999). A susceptibility region for myasthenia gravis extending into the HLA-class I sector telomeric to HLA-C. *Hum Immunol,* 60: 909 –917.

[642] Gilhus, N. E. (2016). Myasthenia Gravis. *N Engl J Med,* 375(26): 2570-2581.

[643] Compston, D. A., Vincent, A., Newsom-Davis, J. and Batchelor, J. R. (1980). Clinical, pathological, HLA antigen and immunological evidence for disease heterogeneity in myasthenia gravis. *Brain,* 103(3):579-601.

# ABOUT THE AUTHORS

*Davide Giacomo Corda*
Department of Clinical
Surgery and Experimental Medicine
University of Sassari
Sassari, Italy
dvcorda@tiscali.it


*Giovanni Andrea Deiana*
Department of Clinical
Surgery and Experimental Medicine
University of Sassari
Sassari, Italy
gdeiana@uniss.it

*Giannina Arru*
Department of Clinical
Surgery and Experimental Medicine
University of Sassari
Sassari, Italy

***Giovanni Masala***
Department of Clinical
Surgery and Experimental Medicine
University of Sassari
Sassari, Italy


***GianPietro Sechi***
Department of Clinical
Surgery and Experimental Medicine
University of Sassari
Sassari, Italy
gpsechi@uniss.it

# INDEX

## U

## V